The Development of Radiation Protection in Diagnostic Radiology

Author:

Stewart C. Bushong
Baylor College of Medicine
Texas Medical Center
Houston, Texas

published by:

18901 Cranwood Parkway, Cleveland, Ohio 44128
A division of The Chemical Rubber Co.

This book represents information obtained from authentic and highly regarded sources. Reprinted material is quoted with permission, and sources are indicated. A wide variety of references are listed. Every reasonable effort has been made to give reliable data and information, but the author and the publisher cannot assume responsibility for the validity of all materials or for the consequences of their use.

Library of Congress Card No. 73-82207
International Standard Book No. 0-87819-120-8

THE AUTHOR

Stewart C. Bushong is Associate Professor of Radiological Science, Baylor College of Medicine, Texas Medical Center, Houston. He received his B.S. degree from the University of Maryland and his M.S. and D.S. degrees from the University of Pittsburgh.

This book originally appeared as an article in *CRC Critical Reviews in Radiological Sciences,* a quarterly journal published by The Chemical Rubber Co. Dr. Donald G. Willhoit, Associate Professor, University of North Carolina, School of Public Health, served as referee for the article.

PREFACE

Almost immediately after their discovery, x-rays were applied as a diagnostic medical tool and during the past seven decades this application has continually expanded in both quantity and quality. No one questions the beneficial role of x-ray diagnosis in medicine, but during recent years concern has increased over the possible harmful effects of x-ray exposure. In order to minimize these harmful effects, radiation protection guides, radiation protective equipment, and radiation control practices have been developed and applied.

This book traces the evolution of these radiation control measures in diagnostic radiology. In the early days of radiology the primary concern for the consequent harmful effects of radiation exposure was directed toward the patient. The extremely high exposures required at that time to produce a diagnostic radiograph often resulted in acute effects such as erythema, desquamation, and epilation. During the second and third decades of this century x-ray equipment was improved to provide increased kV and mA so that high patient exposures were not necessary. Concern for radiation injury shifted from the patient to the radiologist. Radiologists through the 1930's received inordinately high occupational radiation exposures by today's standards and evidence suggested that these exposures resulted in harmful latent effects, particularly hematological disorders. Development of protective devices and procedures has resulted in a marked decrease in occupational radiation exposures accompanied by a removal of fear for latent consequences. It was not uncommon for radiologists in the 1930's to receive 1 rem/week occupational exposure. Currently this is more the order of 500 to 1000 mrem/year. Concern for radiation injury has shifted back to the patient. Although radiation exposures to patients and radiologists are extremely low in today's practice, there is mounting evidence based on epidemiological studies that the use of x-rays in medical diagnosis is responsible for some latent population effects such as an increased incidence of leukemia and cancer. Therefore, continued development of radiation control measures is necessary.

In this volume the history of radiation control in diagnostic radiology is reviewed. Currently recommended radiation protection guides are discussed and radiation protection devices associated with modern x-ray equipment are described. Methodology, techniques, and practices employed in present-day diagnostic radiology that contribute to reduced patient exposure without sacrificing diagnostic information are analyzed.

Stewart C. Bushong
Houston, Texas

TABLE OF CONTENTS

1. INTRODUCTION

In all probability, Wilhelm Conrad Roentgen did not discover x-rays. There is substantial evidence that could be presented in support of any one of a dozen other research scientists as the discoverer of x-radiation.[1,2] Nearly all physics laboratories in the 1880's and early 1890's had the necessary equipment to produce x-rays. Many physicists, working with Crookes's tubes, had noticed that photographic plates secured in light-tight wrappings were often mysteriously fogged. None pursued this phenomenon until Roentgen. For example, in 1890 Professor Arthur W. Goodspeed, a Philadelphia physicist and Roentgen pioneer, exposed coins to what he thought were cathode rays producing a radiograph of the coins. He did not recognize the significance of his experimental results. After Roentgen's discovery five years later, it was apparent that Professor Goodspeed had x-rayed his coins. Roentgen, however, is the recognized father of x-radiation, principally because he was the first to publish about this agent,[3,4] and secondly, because within approximately one year of his first publication, he had conducted investigations that allowed him to describe very thoroughly x-radiation, its physical characteristics, its anticipated medical applications, and its expected harmful biological effects.

As a matter of fact, in the first few years following Roentgen's discovery, the great flurry of research activity that began in an effort to more precisely identify this new agent resulted in numerous radiation injuries, many of them documented. These injuries, however, did not precede demonstrated medical applications of x-radiation. The Roentgenographic files are replete with examples of early medical applications beginning with the famous radiograph of Mrs. Roentgen's hand. By the year 1900, Roentgenography was practiced by many scientists and physicians and had become generally accepted by the medical community as a useful diagnostic tool. This early history of Roentgenology has been the subject of several stimulating books[1,2,5-7] and articles[8-10] that are recommended for the interested reader.

Within a year of Roentgen's discovery, Henri Becquerel discovered radioactivity and ushered in another new era for medicine. Shortly after Becquerel's discovery, radium was being applied to the treatment of cancer. The application to medicine of large quantities of different radioisotopes had to await the development of the atomic technology following World War II. The present rapidly expanding specialty of nuclear medicine can be traced directly through the Curies and Rutherford to Becquerel's discovery.

Radioactive materials and radiation-producing machines are being employed with ever increasing frequency. The beneficial application of these agents is unquestioned. However, the beneficial application of these agents with a proper respect for possible long-term deleterious effects is being questioned ever more closely by physicians and radiation scientists alike[11] in addition to popular magazine writers.[12,13] There is no question but that ionizing radiation in high, short-term doses can result in disastrous biological effects. The precise effects from low-dose, long-term radiation exposure to the individual or to society are unknown. Because of our increasing application of ionizing radiation to man, this lack of knowledge dictates that caution be exercised in each radiological procedure.

Man has occupied this planet for many thousands of years and has evolved his present characteristics while living in an environmental radiation level of approximately 100 mrads per year. Because it has been shown that ionizing radiation can produce genetic mutations, the evolutionary process may be a result of this radiation exposure. Perhaps only a very small percentage of the mutations that have accompanied the evolutionary process are radiation induced; however, few, if any, would rule out completely the role of ionizing radiation in evolution. Therefore, some might say that low-level, long-term radiation exposure appears to be beneficial. On the other hand, there is no question but that high levels of radiation exposure result in not only genetic mutations but also somatic effects. Radiation has been shown to be leukemogenic, carcinogenic, cataractogenic, and a contributor to some non-specific premature aging and death. Undoubtedly, high levels of radiation exposure are harmful.

The approach of responsible radiation scientists today in establishing recommendations for radiation control in diagnostic radiology is to attempt a proper balance between the hazard to patient and radiological worker consistent with the benefits provided. Basically, the guidelines under which we are occupationally exposed to radiation are founded upon the linear, non-threshold hypothesis of dose versus response. Simply put, this hypothesis states that all unnecessary radiation exposure should be avoided since even the smallest dose has a finite probability of producing a deleterious effect. These statements apply equally well to the medical application of ionizing radiation as to other applications of ionizing radiation.

The present increasing concern for restricting exposure to the patient as well as to the operator in diagnostic radiology is based on a number of considerations. Over the past few decades, the increasing application of diagnostic radiology has resulted in an estimated 50% increase in the total radiation exposure to man over previous estimates of the average genetic dose. In other words, having lived for thousands of years in an environment of approximately 100 mrads per year, we are now receiving on the average approximately 150 mrads per year and the bulk of this increase is due to diagnostic applications of x-rays. Table 1 contains approximate values of the average annual genetically significant dose to man. Although the source of highest exposure is naturally occurring, diagnostic medical applications constitute a large and increasing portion of the total radiation exposure. Animal experiments to date would indicate that these diagnostic levels of exposure are not harmful. However, the animal experiments are much too short-term and contain too few experimental subjects; therefore, they are totally inadequate to predict accurately the possible long-term genetic effects of low radiation exposures such as experienced in diagnostic radiology. Another equally serious consideration relates to the suspected increase in late somatic effects following diagnostic levels of x-radiation. It is impossible to demonstrate on an individual basis that some

TABLE 1

Approximate Annual Genetically Significant Radiation Doses to the United States Population from Several Sources

Radiation Source	Radiation Dose (mrem)
Naturally occurring external	80
Naturally occurring internal	20
Diagnostic x-rays	55
Diagnostic Radionuclides	1
Other (luminescence dials, color tv, etc.)	1
Nuclear Power Plant Environs	1

latent effects such as leukemia or cancer are due to prior diagnostic x-ray exposure. However, there seems little doubt that an increase in late somatic effects occurs when those effects are quantitated using epidemiological techniques.

Therefore, we face a situation where our increasing diagnostic application of x-radiation makes it necessary that we do all that is practically possible to maintain the lowest useful levels of exposure consistent with the medical information required. In the following pages, this review will trace our knowledge of radiation effects from early radiation accidents to current experimental findings in an effort to develop an understanding of the basis and foundation for current radiation protection standards and practices in diagnostic radiology. The appropriate recommended standards will be reviewed and a discussion of current practices and procedures relating to exposure reduction in diagnostic radiology will be presented.

2. EARLY X-RAY APPLICATIONS AND INJURIES

Immediately following Roentgen's discovery, it was apparent that this new agent would have a tremendous impact on medical diagnosis. Its therapeutic applications became apparent inadvertently following radiation injuries and possibly were more slowly developed. The tragedy that accompanied the early development of Roentgenology is graphically recorded in the literature by the reports of deformity and death to many Roentgen pioneers.

Professor Roentgen described in his first paper[3] an x-ray of a hand that clearly showed the bony structure and a ring on the ring finger. Immediately following the publication of this report, there was a flurry of activity in physics laboratories throughout the world in an effort to confirm Roentgen's findings and extend his experiments. To produce x-rays was not difficult, for nearly every physics laboratory had the necessary equipment on hand. Crookes's tubes were abundant, induction coils were available, and photographic plates were widespread. Nothing more was needed except for ingenuity. Roentgen's ingenuity was shown by the statements of his first paper which was translated into English within a month of its publication in Germany:[14]

"If the discharge of a fairly large induction coil be made to pass through a Hittorf vacuum tube, or through a Lenard tube, a Crookes's tube, or other similar apparatus, which has been sufficiently exhausted, the tube being covered with thin, black cardboard which fits it with tolerable closeness, and if the whole apparatus be placed in a completely darkened room, there is observed at each discharge, the bright illumination of a paper screen covered with black barium platinocyanide, placed in the vicinity of the induction coil, the fluorescence thus produced being entirely independent of the fact whether the coated or the plain surface is turned towards the discharge tube. This fluorescence is visible even when the paper screen is at a distance of 2 meters from the apparatus.

It is easy to prove that the cause of the fluorescence is received from the discharge apparatus and not from any other point in the conducting circuit.

The most striking feature of this phenomenon is the fact that an active agent here passes through a black cardboard envelope, which is opaque to the visible and the ultraviolet rays of the sun, or of the electric arc; an agent, too, which has the power of producing fluorescence.

The justification for calling by the name "rays" the agent which proceeds from the wall of the discharge apparatus, I derive in part from the entirely regular formation of shadows, which are seen when more or less transparent bodies are brought between the apparatus and the fluorescent screen (or the photographic plate).

I have observed, and in part photographed, many shadow pictures of this kind, the production of which has a particular charm. I possess, for instance, photographs of a shadow of a profile of a door which separates the rooms in which, on one side, the discharge apparatus was placed, on the other the photographic plate; the shadow of the bones of the hand; the shadow of a covered wire wrapped on a wooden spool; of a set of weights enclosed in a box; of a compass in which the magnetic needle is entirely enclosed by metal; of a piece of metal whose lack of homogeneity becomes noticeable by means of the X-rays, etc."

Almost immediately Roentgen ray apparatus found its way into the hospital and into the office of the physician. Its use in medical diagnosis of bony structures was obvious. This is attested to by the fact that in the early days radiographs of the hands were commonplace. Glasser[6] offers examples of nine of these early radiographs and comments that if it had not been for the fascination created by them, x-radiation might have remained for some time simply a tool of the laboratory physicist.

During the year 1896 both the scientific and the lay literature was filled with Roentgenological reports. These reports came so rapidly throughout the world following Roentgen's discovery in December 1895 that it is next to impossible to identify significant first applications.

German physicians were among the first to seize upon Roentgen's discovery and apply it to their own interests. An example of this is shown by the report by Dr. F. Konig[15] that appeared in June 1896:

"On January 29, we made the first Roentgen picture of a patient. It was a boy referred to us by Dr. Tischendorf. The patient had injured his second metacarpal bone of his right hand. The picture was made with an exposure time of four minutes with the photographic plate at a distance of 24 cm from the center of the tube. The extent of the injury could be seen on the plate very clearly. Later we made a large number of pictures within a short time. On February 1, we took a picture of a foreign body, a needle, imbedded in the hand of a girl referred by Dr. Harbordt. And, on February 2, we made the first picture of the teeth by placing films protected from light by a wrapper into the mouth; with Dr. Tischendorf we also made pictures of many other subjects, animals, parts of mummies, etc."

The date of the first diagnostic radiograph in the United States was probably February 3, 1896. This film was made by Edwin Frost in the Physics Laboratory of Dartmouth College and showed a fracture of the left wrist of the patient.

Many of the early Roentgen pioneers, such as Thomas Edison and Professor Goodspeed, were physicists. One American scientist of that period who stands out is Professor Michael Pupin. Professor Pupin has briefly described his early activities with x-rays and claims many "firsts" for which he was not recognized. His comments are characteristic of the activities of many of his contemporaries including the disputed claims.[16]

"No other discovery within my lifetime had ever aroused the interest of the world as did the discovery of the X-rays. Every physicist dropped his own research problems and rushed headlong into the research of the X-rays. The physicists of the United States had paid only small attention to the vacuum tube discharges. To the best of my knowledge and belief I was at that time the only physicist here who had had any laboratory experience with vacuum tube research, and I got it by overtime work in the electrical engineering laboratory of Columbia College. I undertook it because my intercourse with Mueller, the glassblower of Berlin, directed my attention to this field of research, and particularly because I did not see that with the equipment of that laboratory I could do anything else. I decided, as mentioned above, to leave the field to Professor J. J. Thomson, of Cambridge, and to watch his work. When, therefore, Roentgen's discovery was first announced, I was, it seems, better prepared than anybody else in this country to repeat his experiments and succeeded, therefore, sooner than anybody else on this side of the Atlantic. I obtained the first X-ray photograph in America on January 2, 1896, two weeks after the discovery was announced in Germany.

Many interesting stories have been told about the rush to the West during the gold-fever period, caused by the discovery of gold in the far West. The rush into X-ray experimentation was very similar, and I also caught the fever badly. Newspaper reporters and physicians heard of it, and I had to lock myself up in my laboratory, which was in the cellar of President Low's official residence at Columbia College, in order to protect myself from continuous interruptions. The physicians brought all kinds of cripples for the purpose of having their bones photographed or examined by means of the fluorescent screen. The famous surgeon, the late Doctor Bull of New York, sent me a patient with nearly a hundred small shot in his left hand. His name was Prescott Hall Butler, a well-known lawyer of New York, who had met with an accident and received in his hand the full charge of a shotgun. He was in agony; he and I had mutual friends who begged me to make an X-ray photograph of his hand and thus enable Doctor Bull to locate the numerous shot and extract them. The first attempts were unsuccessful, because the patient was too weak and too nervous to stand a photographic exposure of nearly an hour. My good friend, Thomas Edison, had sent me several most excellent fluorescent screens, and by their fluorescence I could see the numerous little shot and so could my patient. The combination of the screen and the eyes was evidently much more sensitive than the photographic plate. I decided to try a combination of Edison's fluorescent screen and the photographic plate. The fluorescent screen was placed on the photographic plate and the patient's hand was placed upon the screen. The X-rays acted upon the screen first and the screen by its fluorescent light acted upon the plate. The combination succeeded, even better than I had expected. A beautiful photograph was obtained with an exposure of a few seconds. The photographic plate showed the numerous shot as if they had been drawn with pen and ink. Doctor Bull operated and extracted every one of them in the course of a short and easy surgical operation. Prescott Hall Butler was well again. That was the first X-ray picture obtained by that process during the first part of February, 1896, and it was also the first surgical operation performed in America under the guidance of an X-ray picture. This process of shortening the time of exposure is now universally used, but nobody gives me any credit for the discovery, although I described it in the journal, *Electricity*, of February 12, 1896, before anybody else had even thought of it. Prescott Hall Butler was much more appreciative and he actually proposed, when other offers to reward me for my efforts were refused, to establish a fellowship for me at the Century Club, the fellowship to entitle me to two toddies daily for the rest of my life. This offer was also refused."

As with so many of the early radiographic applications, Pupin's use of a fluorescent screen-film combination was lost for a time and did not come into general use until many years later.

Roentgen diagonosis was not limited in the early years to high density structures. In June of 1896, J. C. Hemmeter[17] published a paper entitled, Photography of the human stomach by the Roentgen method. By the end of 1896 numerous examples of soft tissue radiographs were available. Early radiographs – the term radiograph having first been suggested by Professor Goodspeed[18] – were of poor quality because of the low x-ray intensities available and the limited penetrability of these x-rays. X-ray exposures lasting one hour or more were not uncommon, and these created rather severe hardships on the patient and the examiner alike. The early medical adventures were usually accomplished by the physician bringing his patient to the laboratory of the physicist, since the physicist had the necessary equipment and experience. The demand for x-ray examination became so great that before the year 1896 was through, each major city in the U. S. had at least one centralized x-ray examination laboratory. Slowly, as the physicians became familiar with the procedure, they began to acquire their own equipment and perform their own examinations. With the publication of radiographs of internal organs, vascular structures, and the skull, x-ray diagnosis became as important a tool for the internist as for the surgeon.

One of the most active and thorough experimenters with x-rays during the first few years after their discovery was Thomas A. Edison. According to his assistant, Mr. Meadowcraft:[18]

> "Mr. Edison was the first to recognize the importance of the cable announcement of Dr. Roentgen's discovery. The same day he started to make the apparatus and had it finished and working on the next day. Three of the metropolitan dailies heard of it, and for three weeks, more than 20 newspaper reporters were stationed at the Laboratory, the work going on nights, days and Sundays."

Edison's feverish activity caused him to complain on several occasions early in 1896 of severe pain and soreness in his eyes.[19] He did not attribute these symptoms to radiation, but others subsequently did. One of Edison's assistants, Clarence M. Dally, became the first radiation fatality in this country. Dally suffered from a severe radiodermatitis and died in 1904. As Meadowcraft has noted:[18]

> "About a year after the exhibition (the 1896 electrical exhibition at the Grand Central Palace in New York), Mr. Edison's assistant, Clarence Dally, commenced to lose his whiskers and his hands became inflamed. Several surgical operations followed, including the amputation of his arm, etc. But, he soon died of X-ray cancer, being the first known victim of the X-ray."

Because of the loss of his friend and assistant, Edison turned his attention from x-radiation to other areas of investigation. He had discovered calcium tungstate as a replacement for barium-platinocyanide in fluorescent screens and had developed the fluoroscope. Surely, radiology would have made much more rapid progress in the early years had Edison stayed active in that field. Another tragic loss of an x-ray pioneer was Walter J. Dodd, a pharmacist and photographer at the Massachusetts General Hospital. In November 1896, Dodd developed a severe radiodermatitis that was described,[18] "as though his hands and face had been scalded." Dodd underwent a series of skin grafts in an effort to control his painful condition. Walsh[20] described harmful radiation effects on internal organs in 1897. Colwell and Russ[21] have reported an autobiographical note found in the October 29, 1896, issue of *Nature*. Their report clearly and characteristically describes the superficial injuries that were becoming commonplace:

> "The narrator was employed as an X-ray demonstrator at the Indian Exhibition held in London in 1896. Commencing work in May 1896, he was exposed daily for several hours. For two or three weeks he suffered no apparent injury, but at the end of this time there appeared 'little blisters of dark colour under the skin' of the right hand. These gradually became very irritating, and the skin red and inflamed. The patient tried *lotio plumbi* as a sedative, but with only very temporary benefit; the only measure seeming to afford any ease was to plunge the hand into the coldest water obtainable. The smarting became so bad that he was on the point of giving up work, when he was recommended to try some proprietary ointment. The irritation disappeared at once and the use of the ointment was continued during the remainder of the patient's engagement.
>
> The skin of the fingers became yellow, parchment-like, and finally peeled. In the middle of July there was swelling of the finger-tips and the skin showed signs of marked tension; at the same time the fingernails began to give trouble. Pain and discomfort became severe, and were only partially relieved when a copious, colourless, and unpleasant-smelling discharge came from beneath the nails. This continued until the nails were shed. The swelling of the finger-tips decreased, but pain was severe when separation of the old and newly formed nails took place. The old nails became hard and black, and the fingers had to be bandaged for six weeks. In mid-August the left hand began to show the first signs of trouble. The skin of the right hand was shed for the third time, and the patient decided to try the effect of lanoline. This was well rubbed in and the hands protected with ordinary kid gloves, which became soaked with the grease. From this time until the date of writing the note (Oct. 17) there was no further peeling and the patient was much more comfortable. For six weeks he was unable to hold a pen, which was only possible after the nails were shed and the tension of the finger-tips relieved. Four nails were lost on the right hand, two on the left, while two more were expected to be shed at any time."

These early reports were only the beginning. As Stone relates,[22] by the end of 1896, 23 cases of severe x-ray dermatitis had appeared in the literature. According to Rolleston,[23] between the years 1911 and 1914, three reviews appeared that collectively identified 198 cases of radiation induced malignancy and that reported 54 deaths from these causes. Still later, in 1934, Colwell and Russ[21] reported that by that time more than 200 radiologists had died of radiation induced malignant disease.

Because of the increasing frequency of reports telling of radiation injury, a number of early x-ray users recognized the possible therapeutic value of this agent. Perhaps the first therapeutic application of x-rays occurred on January 29, 1896.[22] Emil Herman Grubbe, who was experimenting with and manufacturing Crookes's tubes in Chicago, developed a radiodermatitis in December 1895. He and Dr. J. E. Gilman, a colleague, reasoned that if x-rays could produce that type of reaction on normal tissue, they should be tried therapeutically. Mr. Grubbe treated one of Dr. Gilman's patients, Mrs. Rose Lee, with x-radiation for an advanced carcinoma of the left breast. Grubbe, by the way, was probably the first radiation casualty. He lived until 1960 but was subjected to many disfiguring operations for control of radiation induced neoplasm. He described some of his experiences, as well as those of others in a book entitled, *X-Ray Treatment: Its Origin, Birth and Early History.*[24] He brought to this work first-hand knowledge, both as a practitioner and as a patient.

Dr. Freund, a Vienna radiologist, treated a hairy pigmented birthmark on a young child.[6] The treatment consisted of two-hour irradiations for a period of ten days, and Freund subsequently reported success in removing the ugly birthmark. A French physician, Dr. Despeignes, treated with some success an oral carcinoma and a carcinoma of the stomach in early 1896.[6]

The early x-ray injuries were overt, obvious, and acutely manifested, and it was on the basis of this tragic experience that equipment was designed for safety and that radiation protection standards were established. When considering the brilliant achievements of these early x-ray pioneers, we would do well to keep in mind the martyrdom that befell many of them.

3. EARLY RADIATION PROTECTION STANDARDS AND THEIR BASIS

In view of the large number of early radiation injuries and the manner in which they were reported in the scientific and lay press, it is surprising that so many years passed before formal recommendations restricting the use of ionizing radiation were made. Kathren,[25] in his summary of radiation protection actions in the first decade of Roentgenology, lists no less than 17 "X-Ray Protection Efforts" through 1903. Nevertheless, it was 1915 before the first radiation protection guides were issued by a scientific society or committee.[26] Some individual radiation scientists had been promoting certain restrictions for a number of years, but, as Kathren[25] points out, these did not receive particular attention at that time.

Radiation protection standards have undergone a continual scrutinization since that time, the result being a periodic reduction in the amount of radiation considered acceptable as an occupational risk. The history of the establishment of radiation protection standards has been and continues to be fraught with two dilemmas. The first is a precise definition, meaning, and understanding for the upper limit of radiation exposure that is deemed acceptable—the quantity now identified as the Maximum Permissible Dose (MPD). The second is the identification of an underlying radiation dose-response relationship from which an MPD can be derived. These problems faced the early radiation scientists, and they continue to plague us today.

3.1 Recommendations of Pioneer Radiation Scientists

Perhaps the one man, more than any other, who can be identified as the Father of Radiation Protection was Dr. William Rollins, a Boston dentist. Stone,[27] while analyzing some early reports on radiation protection, indicated, "The first reference I found to a tolerance dose was by Rollins in 1902." Rollins did, in fact, specify the maximum exposure level that should be allowed to emanate from any x-ray tube. In 1902 he proposed a simple test to determine the adequacy of x-ray tube shielding.[28] He strenuously supported shielding in addition to the glass envelope of the tube, and the basis for his protective standard was: ". . .expose a photographic plate in contact with the outside of the case. If the plate does not fog in seven minutes, the coating is sufficient." Under our current system of measure, this leakage radiation would result in a personnel exposure of perhaps 10 to 20 R per day of soft radiation.

This recommendation is not the main reason that Rollins is so prominently identified in the literature associated with radiation protection standards. During the period from 1900 to approximately 1904, Rollins published numerous notes in scientific journals warning of the hazards of overexposure to x-radiation. He constantly urged radiologists to use the smallest exposure possible in all procedures. He made numerous recommendations for methods of reducing exposure to operators and patients which were the forerunners of our current cardinal principles of radiation protection, i.e., time, distance, and shielding. Many of his early pronouncements were published in the *Boston Medical and Surgical Journal*, in the *Electrical Review*, and in a book that he authored and privately published in 1904.[29] Some of his more profound statements have been cited by Brecher[2] and are repeated here because of their farsighted implications.

> "A pregnant guinea pig was placed in a closed metallic chamber hung by silk cords within a closed metallic chamber connected with the ground. This is the same arrangement for excluding electricity as a factor in the results which was used in the other experiments. The source of the X-light was outside. Exposure to X-light killed the foetus. . .This notice published because pregnant women are being exposed to X-light to determine the size of the pelvis or to examine the condition of the foetus.
>
> It seems hopeless even at this late date, when so much is known about the possibilities of injuries from X-light, to get others to use an X-light tube entirely enclosed in a nonradiable case because of the supposed inconvenience of a box.
>
> A tube case needs an instantaneous shutter. . ., then by pressing and releasing a bulb, the shutter is opened, the plate exposed to the most suitable radiation and the shutter closed.
>
> Apparatus which requires a less distance than one meter between the (tube) and the photographic plate is not satisfactory.
>
> No X-light should strike a patient except the smallest beam which will cover the area to be examined, photographed or treated.
>
> Selective filters should be used to strain out undesirable radiations.
>
> The opening in the diaphragm plate in the X-light tube box should be rectangular for diagnostic and photographic work because this is the form of the fluorescent screen and photographic plate. If we use a round opening, the section of the cone of X-light escaping from the tube box is a circle . . .While the patient will be illuminated by the whole cone, it is evident that the only part of the illumination which will be useful will be that included in the rectangular area."

Thus, Dr. Rollins was several years and, in some cases, several decades ahead of his time in proposing radiation control procedures. For example, although rectangular collimation is currently employed in nearly all diagnostic radiological procedures, this equipment adaptation has not found its way into dental Roentgenography. As Winkler[30] points out, most dental x-ray machines are still fitted with round collimators even though the x-ray film is rectangular. Rectangular collimation would result in less than half the patient dose as round collimation, according to Winkler's data.[30]

That Dr. Rollins' recommendations were not immediately heeded is understandable by a perusal of the early literature. Many radiologists, in spite of the radiation injury reports, were against strict radiation control. In 1907 Dr. W. S. Laurence remarked in a published discussion session:[31]

> "It seems to me that there is no such thing as absolute protection unless you go to an adjoining town and operate by telephone. Then the question is how much protection is it wise to take. I believe that there is a point of tolerance in the human body to the X-ray as there is to other injurious agents. For that reason, I do not believe that it is necessary to clothe ourselves in armor or to leave the city and operate our apparatus from a distance. To work without protection is foolhardy and inexcusable, but I believe attempts at absolute protection have been carried to rather absurd extremes."

Another participant in that discussion was Dr. E. W. Caldwell who stated:[32]

> "Undoubtedly we should avoid all unnecessary exposure, but I believe that there has been undue alarm over the dangers of the secondary rays and that we are not warranted in sacrificing facilities for observing and controlling our tube in order to exclude the last vestige of secondary radiation. I think that a screen which will protect the photographic plate so well that only a little fog is shown on development after two or three weeks' exposure in the position occupied by the operator is practically safe even if all the secondary rays are not excluded."

This same Dr. Caldwell is memorialized annually by lecture title at the American Roentgen Ray Society meeting and by the award of the Caldwell Medal to an outstanding radiologist. Dr. Caldwell died in 1918 from radiation injuries.

3.2 Recommendations of Radiological Societies

Between the time of Rollins and Dr. Sidney Russ, very little can be found other than occasional remarks relative to minimizing radiation exposure. In the period of 1914 to 1916, Dr. Russ, a member of the British Roentgen Society, urged that society take positive action in formulating radiation protection standards.[33,34] His concern was voiced as follows:[33]

> "Now the question arises as to whether the dangers of X-rays are becoming more or less formidable and I think we have to recognize that they are becoming less formidable on account of the wider appreciation of the dangers which exist, but we ought also to remember on the other hand, that greater output of radiation now

available makes it necessary to take still more stringent precautions. The output of radiation from the modern X-ray apparatus is now just about 50 times what it was 10 years ago with the apparatus then available. Are our precautionary measures running parallel with this enormous increase in output?"

Dr. Russ had been urging the Roentgen Society to take a formal stand for stricter radiation protection guidance and to support restrictions on the operation of x-ray apparatus to qualified operators. He also called for established standards for the manufacturer and for the inspection of x-ray apparatus. His efforts resulted in two notable pronouncements. In November of 1915 the Roentgen Society issued "Recommendations for the Protection of X–Ray Operators."[35] Similarly, in July 1916, the Society passed a strongly worded resolution with the intention of influencing government regulations.[36] Each of these pronouncements was concerned with adequate equipment design and proper equipment operation. Neither, however, attempted to describe a maximum permissible dose for either the patient or the operator.

Following Russ's activities, there was a period of several years during which the radiology literature continued to be filled with papers describing the deleterious effects of ionizing radiation. There was a scarcity, however, of radiation protection papers. In 1920 the American Roentgen Ray Society (ARRS) and in 1921 the British Roentgen Society, respectively, formed committees for the purpose of investigating further requirements for radiation protection. The British X-Ray and Radium Protection Committee issued its first reports in 1921,[37,38] while the ARRS Committee issued its first reports in September 1922.[39] These reports were quite general in nature and indicated only the need for rudimentary radiation protection measures and the exercise of prudence in the application of x-radiation. Neither committee specified, or even alluded to, a maximum permissible dose. Table 2 is a restatement of the initial

TABLE 2

Preliminary Statements from the 1921 Report of the British X-ray and Radium Protection Committee

The danger of over-exposure to x-ray and radium can be avoided by the provision of efficient protection and suitable working conditions.

The known effects on the operator to be guarded against are:

1. Visible injuries to the superficial tissues which may result in permanent damage.
2. Derangements of internal organs and changes in the blood. These are especially important as their early manifestation is often unrecognized.

GENERAL RECOMMENDATIONS

It is the duty of those in charge of x-ray and radium departments to ensure efficient protection and suitable working conditions for the *personnel*.

The following precautions are recommended:

1. Not more than seven working hours a day.
2. Sundays and two half-days off duty each week, to be spent as much as possible out of doors.
3. An annual holiday of one month or two separate fortnights. Sisters and nurses, employed as whole-time workers in x-ray and radium departments, should not be called upon for any other hospital service.

PROTECTIVE MEASURES

It cannot be insisted upon too strongly that a primary precaution in all x-ray work is to surround the x-ray bulb itself as completely as possible with adequate protective material, except for an aperture as small as possible for the work in hand.

The protective measures recommended are dealt with under the following sections:

I. X-rays for diagnostic purposes.
II. X-rays for superficial therapy.
III. X-rays for deep therapy.
IV. X-rays for industrial and research purposes.
V. Electrical precautions in x-ray departments.
VI. Ventilation of x-ray departments.
VII. Radium therapy.

It must be clearly understood that the protective measures recommended for these various purposes are not necessarily interchangeable; for instance, to use for deep therapy the measures intended for superficial therapy would probably subject the worker to serious injury.

portion of the 1921 British report.[37] For some reason particular emphasis was placed on leisure time for radiological workers (see General Recommendation Number 3, Table 2). As Desjardins stated in the last sentence of a paper entitled Protection against Radiation:[40] "The cultivation of an outdoor hobby is of special importance to all persons exposed to radiation."

These committees were stimulated to action more from the outside than from their own membership. The first individual to recommend a numerical maximum exposure level based on empirical evidence was probably Dr. Arthur Mutscheller, an American physicist. In a 1925 paper,[41] Mutscheller expressed his concern in the first sentence:

> "The results of inadequate protection against the harmful effects of overdosage with Roentgen rays that has been reported in the past, are so appalling that a search for protective standards is one of the most important problems in Roentgen ray physics and Roentgen ray biology."

Dr. Mutscheller measured the radiation environment at a number of Roentgen ray laboratories in the New York City area, and he observed that the radiologists and technicians at these facilities had not suffered any ill effects. On this basis, he concluded that the exposure levels present in these laboratories were safe and, therefore, constituted a recommended maximum exposure level. His conclusion was stated as follows:[41]

> "Thus, it seems that under present conditions and standards accepted at present, it is entirely safe if an operator does not receive every thirty days a dose exceeding 1/100th of an erythema dose, and from the present status of our knowledge, this seems to be the tolerance dose for all conditions of operating Roentgen ray tubes for roentgenography, roentgenoscopy, and therapy."

Assuming that the erythema dose was 400 rads, Mutscheller's recommendations would be equal to about 200 mrad/day. In 1928 Mutscheller reviewed measurements made by a number of other investigators and additional measurements that he himself made and arrived at the same permissible dose level that he had determined in 1925–1/100th of an erythema dose per month.[42] In 1934 Mutscheller again reiterated his recommendation of 1/100th of an erythema dose per month.[43] However, at this time, he did point out that it had been demonstrated that the erythema dose was very much dependent upon the radiation energy. Consequently, he expressed his permissible dose in terms of the newly agreed upon unit, the Roentgen. Estimates of his earlier recommendations, based on a superficial type radiation, resulted in a permissible dose of 3.4 rad per month, while higher energy radiation would allow up to 7.5 rad per month.

Mutscheller, though his publications were well received, was not the only radiation scientist to have a noticeable effect on these early radiation protection standards. In fact, during the 1920's and 1930's, there were many whose writings influenced the decisions of the scientific advisory committees. In 1922, Pfahler delivered his presidential address to the Seventh Annual Meeting of the American Radium Society and it was entitled, Protection in Radiology.[44] He offered six general recommendations ranging from protective devices to protection by legal counsel, because as he put it:

> "There have been too many martyrs to the science of radiology. The loss of each one of these has produced the gap in the rank of leaders and has delayed progress. There are still among us many who are being damaged physically."

By the early 1920's it was becoming apparent that, although then currently employed protection practices were effective in reducing the incidence of radiodermatitis and its sometimes hideous latent sequelae, a possibly more insidious danger was present. At this time, radiation induced cancer was an accepted reality since over 100 cases had been documented. Radiation induced blood disorders were becoming more and more apparent, and reports began to appear suggesting that radiation was responsible for some deaths attributed to aplastic anemia and lymphocytic leukemia. Many investigators, and Pfahler was prominent among them, reported on observed abnormalities in the peripheral blood and bone marrow of radiation workers. For example, Occupational Hazards of the Radiologist with Special Reference to Changes in the Blood was a lecture delivered before the Radiological Society of North America in 1924 by Drs. Carman and Miller.[45] They documented six deaths from aplastic anemia and three relating radiation with lymphocytic leukemia. Two years later Carman died of gastric carcinoma. Other reports[46-49] in the early 1920's implicated x-radiation exposure with subsequent blood disorders.

In his 1930 critical review entitled, The Harmful Effects of Irradiation, Dr. Humphrey Rolleston presented perhaps the first authoritative study of human radiobiology.[23] He very carefully discussed the literature available to that time relative to the effects of radiation on normal and malignant tissue and to damage to the irradiated fetus following maternal irradiation. He also described the acute radiation syndrome in man as well as chronic systemic and local damage. His review, as the work of Pfahler[44] before him, was prompted by the continuing reports of death by radiation induced malignancy.

Barklay and Cox[49] did not agree with some of these early concerns and in 1928 published their evaluation of the radiation protection situation, along with the recommendation of a numerical permissible dose.

"The effect on the blood of massive doses of high voltage Roentgen radiation and of radium, even in small quantities, is quite definite, but we have considerable doubt as to whether the comparatively infintesimal quantities of radiation to which the Roentgen ray worker is exposed in the course of his diagnostic work, can have any influence in producing serious changes in the blood. We are therefore of the opinion that the very rare cases of blood changes in Roentgen ray workers are due to some other influence. . . ."

Still, Barklay and Cox[49] arrived at some definite recommendations. They measured the average exposure received by two radiation workers who had not suffered any ill effects following long-term employment. The exposures were measured in unit skin dose (USD), or more commonly identified as skin erythema dose (SED), and quite arbitrarily (as they readily admitted) recommended a level of 0.00028 SED per day as a safe level of radiation exposure for unlimited working conditions.

Other radiation protection guides had been proposed on the basis of experience with radium exposure.[50-53] Failla[50] suggested a value of 0.001 erythema dose per month "as the tolerance intensity of gamma rays to which a technician can be exposed continuously for a number of years." According to Failla's measurements, this permissible dose would be equal to 600 mrads per month. Sievert[54] had previously suggested ten times that level and the Dutch Board of Health[55] officially adopted a much lower level.

Cilley et al.[56-60] in a series of papers dealing with radiation protection in fluoroscopy reported on the use of ionization chambers to study the exposure to examining radiologists during various procedures. Their conclusions and recommendations were general in nature, and they avoided the suggestion of a permissible dose. It is interesting, however, that even though they measured rather high exposures to radiological personnel, they could yet conclude:[59]

"In a routine examination in both the stomach and the colon, the examiners wear no protective devices whatsoever. In spite of the fact that these examiners have sustained multiples of the tolerance dose with impunity, we nevertheless advocate no change whatever in the accepted method of protection."

These reports were generated at the Mayo Clinic, and it was not until years later that leaded protective apparel was required at that facility.

Turnbull and Leddy[61] were possibly the first radiologists to express concern for the patient in diagnostic procedures. They measured the exposure per film to patients for various types of examinations and divided this value into their assumed SED of 375 Roentgens. The quotients were tabulated and presented in the form of a maximum number of films per examination that should be allowed. Exceeding this number might cause the patient to receive an erythema dose.

The significant milestones in early attempts to formulate radiation protection guides are listed in Table 3. The dates of the various recommendations and their authors are listed first, followed by the appropriate reference number. The last two columns contain the respective recommendations and this author's estimate of the dose in mrads per day. The table contains recommendations as recent as 1948. A later discussion will relate these to present day standards.

3.3 Recommendations of a Maximum Permissible Dose

Perhaps the single most important event in the early years of radiation protection was the establishment of the unit for radiation exposure intensity, i.e., the roentgen. Up to this time, the most widely used and accepted unit of radiation dose was the skin erythema dose (SED), a very imprecise measure. Just as imprecise was the use of pastille tiles. as cited by Cumberbatch[62] for example: "It was arranged that no portion of red marrow should receive more than one Sabouraud pastille dose (tint B) each month." Early attempts with the use of film for radiation monitoring were equally crude. Collear[63] reports that when films

TABLE 3

Significant Milestones in the Establishment of Early Radiation Protection Guides

Date	Recommended by	Ref.	Recommendation	Approximate Daily Dose (mrads)
1902	Rollins	28	Limited by fogging of a photographic plate following seven minute contact exposure.	10,000
1915	Russ	34	Prompted British Roentgen Society to action.	
1915	British Roentgen Society	36	The need for lead shielding of tube but no numerical exposure levels.	
1921	British X-ray and Radium Protection Committee	37	Described general methods to reduce exposure.	
1922	American Roentgen Ray Society	39	Described general methods to reduce exposure.	
1925	Mutscheller	41	"It is entirely safe if an operator does not receive every thirty days a dose exceeding 1/100 of an erythema dose."	200
1925	Sievert	54	10% of an SED per year	200
1926	Dutch Board of Health	55	One SED per 90,000 working hours	40
1928	International Committee on X-ray and Radium Protection	64	Formation of committee and issuance of general rules.	
1928	Barclay and Cox	49	0.00028 of an SED per day	175
1928	Kaye	73	1/1000 of an SED within 5 days	150
1931	Advisory Committee on X-ray and Radium Protection of the United States	70	Limit exposure to 0.2 R per day	200

Date	Recommended by	Ref.	Recommendation	Approximate Daily Dose (mrads)
1931	Wintz and Rump	74	10^{-5} R/sec for 8 hours.	250
1932	Failla	50	1/1000 of an SED per month	30
1934	International Committee on X-ray and Radium Protection	68	Limit exposure to 0.2 R per day but no conceptual descriptions.	200
1934	Advisory Committee on X-ray and Radium Protection of the United States	71	5 R/day permissible for the hands.	
1935	Cilley et al.	57	250 mR/day was assumed to be safe.	250
1936	Advisory Committee on X-ray and Radium Protection of the United States	72	0.1 R per day	100
1937	International Committee on X-ray and Radium Protection	69	Official acceptance and definition of the unit Roentgen.	
1941	Taylor	75	0.02 R per day	20
1943	Patterson	76	200 mR/per day is acceptable.	200
1943	Parker	77	4 R per week has been shown to cause injury and therefore is an unsafe level.	
1948	Parker	77	100 mR per day is too restrictive.	

SED = Skin erythema dose

were used for this purpose, protective measures were only recommended when a "density that would obscure the readability of newspaper print" was produced.

At the Second International Congress of Radiology (ICR) in Stockholm some physicists in attendance were appointed to an international committee to study appropriate units of x-radiation. They adopted as standard nomenclature the Roentgen, and this unit was accepted by the ICR in 1928.[64,65] Basically, the Roentgen was defined as that quantity of X or γ radiation which would produce in a specified volume of air at standard temperature and pressure a precise quantity of ionization. The wording of the definition was changed several times during subsequent years, without altering the basic measure and understanding of the unit. Certain refinements in the definition, with particular relevance to radiation quality, were incorporated until, finally, at the Fifth International Congress of Radiology in 1937, the Roentgen was officially agreed upon for all existing x-rays and radium gamma rays. The Roentgen was defined at that Congress in the following words:[66]

> "The international unit of quantity or dose of X-rays or gamma-rays shall be called The Roentgen, and shall be designated by the symbol, r. The Roentgen shall be the quantity of X or gamma radiation, such that the associated corpuscular emission per 0.001293 gm of air produced, in air, ions carrying 1 esu of quantity of electricity of either sign."

Since that time, the wording of this definition has been changed, but the concept and value remain the same.

The committee established by the Second International Congress of Radiology to define the Roentgen became the International Committee on X-Ray and Radium Protection (ICRP). The formation of the committee had been suggested at the First International Congress of Radiology in 1925, but there was apparently considerable disagreement over the advisability and structure of such a committee. The first report of this committee was issued in 1928,[64] and was rather brief. It contained primarily information of an organizational nature, with the statement of a few basic general principles of radiation protection. There was no mention of a permissible dose. A subsequent committee report in 1931[67] likewise contained no definitive radiation protection guidelines. Two later reports issued by this committee in 1934[68] and 1938[69] did recommend that occupationally exposed persons not receive more than 0.2 R per day.

This was not the first recommendation for a maximum permissible dose. In 1929 the National Bureau of Standards (NBS) helped to establish the U. S. Advisory Committee for X-Ray and Radium Protection, later to be known as the National Council on Radiation Protection and Measurement (NCRP). The first report of this committee was issued in 1931 in the form of NBS Handbook No. 15,[70] entitled, X-Ray Protection. In this report they recommended that the exposure level of 0.2 R per day not be exceeded. This, of course, was the same level of exposure that was adopted three years later by the ICRP. In 1934 the NCRP issued some additional recommendations in the form of NBS Handbook No. 18[71] entitled Radium Protection. These recommendations reaffirmed the use of 0.2 R per day for the exposure of the whole body and they also established a maximum permissible dose of 5 R per day to the hands. This recommendation was primarily out of regard for those individuals working with radium.

In 1936 the NCRP issued revised recommendations for radiation protection in the form of NBS Handbook No. 20.[72] In this handbook they discussed for the first time the conceptual meaning of a maximum dose to which one might be exposed and established the maximum permissible dose (MPD) for occupationally exposed individuals at the lower value of 0.1 R per week. This MPD level remained the standard for the next 12 years.

The recommended maximum levels of exposure issued by the various scientific committees and societies were established primarily on the basis of radiation injuries reported by a number of investigators during that period of time. Laboratory experimental evidence was generated in quantity and was referred to extensively by the individuals calling for protective standards. Both the human and animal data, however, were consistently concerned with high dose, acute response conditions. The recommendations based on these data, therefore, assumed a threshold response by humans to radiation; i.e., the early levels of permissible exposure were deemed entirely safe. Only exposures in excess of these levels would result in injury. Such is not the foundation for current standards.

4. DEVELOPMENTS IN RADIOGRAPHIC EQUIPMENT

The evolution of modern Roentgenographic equipment has been prompted by the need for increased radiographic quality with reduced radiation exposure. The delivery of more intense radiation from a smaller focal spot in a shorter period of time has been the design requirement. Initially, it was not uncommon for an examination to require an exposure time of several hours in order to get an acceptable film. This was due to both the low voltage and the low tube current available at that time. Almost simultaneously, improvements were made in the x-ray tube design and the voltage generating apparatus so that by 1920 films were taken in less than one second at voltages exceeding 100 kVp. The year 1921 saw the introduction of the first 200 kVp therapy unit[22] and that type of apparatus has evolved through improved equipment of the orthovoltage energy range into the megavoltage range of today's linear accelerators and betatrons. Although improvements are still being made in the basic Roentgenographic equipment today, much of the developmental activity is focused on ancillary and auxiliary apparatus. Unquestionably, developments within the first two decades of Roentgenology were the most significant.

4.1 The X-Ray Tube

Sir William Crookes, like so many of his contemporaries, was a self-taught genius from a humble background. From the 1870's on, Crookes pursued with great vigor his study of the passage of electricity through an evacuated tube. He manufactured many of these evacuated tubes, so many, in fact, that nearly all vacuum tubes of that era became known as Crookes's tubes. In fact, the Crookes's tube was so ubiquitous that, as Crane[9] has put it:

> "Crookes's tubes and induction coils to run them were added to the equipment of high schools and colleges throughout the world. Many hundreds of trained workers operated and lectured upon Crookes's tubes. Tubes varied but a majority must have produced X-rays, yet no one discovered them."

At the time of his discovery, Roentgen was experimenting with a Hittorf-Crookes's tube and a Ruhmkorff induction coil. The Hittorf-Crookes's tube was one of the many modifications of Crookes's tube. It consisted of a nearly evacuated glass envelope with two electrodes separated by several centimeters. With Roentgen's apparatus the x-rays were actually produced by the interaction of the cathode rays (electrons) with the glass envelope. Although the Crookes's tube was identified as a vacuum tube, it was, in fact, a gas tube in present day terminology because the level of vacuum was very critical to its continued operation. In the type of tube that Roentgen used, the rarefied gas was partially ionized and the positive ions would interact with the cathode, liberating electrons from its surface. If the potential difference across the tube were sufficient, the electrons would overshoot the anode and interact with the glass envelope at the end of the tube. The earliest modification to this Crookes's tube came within a year of Roentgen's discovery. The small anode was replaced with a larger, and sometimes concave, metal anode. This resulted in a higher intensity of radiation and a more focused x-ray beam.

The early tubes were very difficult to operate because the level of vacuum had to be precisely controlled. Many types of vacuum regulators were devised—the most successful being the introduction into the tube of a suitable chemical that would emit a gas when heated. The heating was done by an electrical current through an auxiliary circuit that was most often regulated by hand. These tubes were known as "hand regulated tubes."

The evolution of the Crookes's tube did not change much in the next few years. Kaye[78] demonstrated in 1908 that metals of high atomic weight were more efficient than lower atomic weight materials in producing x-rays. In 1913 the Coolidge hot cathode high vacuum tube was unveiled after clinical testing showed it to be far superior to the Crookes's tube.[79] This development transformed a rather stagnant scientific field into a rapidly progressing and expanding medical specialty. The Coolidge tube is essentially the type of x-ray tube in use today. Naturally, it has seen many improvements, some of which can be traced directly to Coolidge. For example, he developed various successive methods for increasing the cooling capacity of his x-ray tube—from air-cooled to water-cooled to an oil immersion tube which he described in 1920. Heat dissipation was, and still is, a major design obstacle in x-ray tubes, and with each of Coolidge's advancements, he was able to increase the voltage and the tube current while, at the same time, decreasing the focal spot size. Coolidge developed the tungsten anode after demonstrating that tungsten could be made ductile in a cold state. It had previously been considered a brittle non-workable metal. It was Coolidge who, when searching for a better target material than aluminum or platinum, stated the four necessary properties of a target that are always reiterated in today's textbooks.

1. High specific weight (atomic number)
2. High melting point
3. High heat conductivity
4. Low vapor pressure

These developments were revolutionary for the practicing radiologist. The Coolidge tube was not only much more reliable than the Crookes's tube, but also this added reliability, and capability, allowed the physician to conduct more examinations, more completely, and to generate considerably more information from already familiar procedures. Vivid contemporary descriptions and references are contained in the volumes written by Kaye.[5,80]

Dr. Lewis G. Cole[81] was the radiologist who first evaluated the Coolidge tube clinically. He called it "undoubtedly the most important contribution to Roentgenology since the birth of that science." Dr. Cole's appraisal has proved correct. The Coolidge tube immediately became standard equipment and has continued in that capacity.

4.2 High Voltage Apparatus

During Roentgen's time there were only two reliable sources of high voltage with which to operate the x-ray tube–an induction coil, or a static machine. Roentgen, as has been noted, used an induction coil, which consisted essentially of two coils of wire, a primary and a secondary, wound around the same iron core. For every turn of the primary, there were hundreds of turns on the secondary; thus, the coil acted as a step up transformer. The induction coil had one major drawback, however. It operated on alternating current, and only direct current was available; therefore, a device known as an interrupter was necessary.

Static machines were used by many early Roentgen workers. These machines were usually large concentric drums, one or both of which could rotate such that a high voltage potential could be produced by means of friction. The drums were made of glass, hard rubber, or mica, and direct current was produced making the static machine very well suited for x-ray tubes.

As experience was gained with each type of voltage generating device, it became obvious that despite its deficiencies, the induction coil was far more suited for x-ray generation than the static machine. Jerman[82] stated, "By 1912, the sale of the static machine for the production of X-rays was a rare event." The use of induction coils was limited by the degree of synchronism provided by the interrupter. Many types of interrupters were developed, none with completely satisfying results.

When the Snook interruptless transformer was introduced in 1907,[2] the induction coil and the static machine became obsolete. The Snook device was more than a transformer–it was an entire power supply. The heart of the device was the closed core transformer, with minimum magnetic leakage, which was sealed in an oil-filled tank. Snook developed several methods for rectifying the output of this transformer and included a rheostat for controlling the output current and a switch for varying the output voltage. Snook's design is basically unchanged today, and it was much ahead of its time because the currents and voltages that were possible with the Snook device far exceeded the capability of the Crookes's tubes. Not until the Snook interrupterless transformer was matched with the Coolidge tube were respectable capabilities of current and voltage realized.

Radiation was not the only hazard to which early radiologists were exposed. Electrical shock was a severe hazard to both patient and operator because of exposed and improperly insulated high voltage lines. Shearer[83] described 19 electrical accidents, some of which were fatal. He listed the causes of each and suggested simple tests to determine whether or not installations were "reasonably safe." For example, one of his recommendations stated:

> "Place a metal plate or water in a metal pail on the floor. Insulate a milliammeter and connect one terminal to the plate or the water, fasten a piece of flexible wire to a dry stick about three or four feet long and approach it to various parts of the apparatus when in operation. The spark distance and the milliampere reading will give a fairly good idea of the danger if the body replaced the wire."

4.3 The Potter-Bucky Grid

Radiographic quality, by today's standards, was abominable to the early radiologists because of image blurring. Coolidge thought that he had overcome this with his smaller focal spot tubes and, indeed, some improvement was noted. But, as Coolidge himself was able to demonstrate, when he x-rayed a lead straight edge, there was still a sizable penumbra which he thought was due to an inadequacy in his tube.

Three researchers,[2] however, working independently, recognized that the blurring was due to scattered radiation generated within the patient. Dr. Gustav Bucky, a German radiologist, Dr. Eugene W. Caldwell, who has previously been mentioned,[32] and Dr. Hollis E. Potter worked on the problem independently and arrived at essentially the same solution. The solution required the insertion of some device between the patient and the x-ray plate to limit scatter radiation. Initial attempts involved crossed grids, but these resulted in too many shadows on the film. Later pursuits followed research leading from a moving crossed grid to a moving parallel grid, and this was a considerable improvement. The ultimate device became known as the Potter-Bucky Diaphragm or grid and is one of the basic tools of modern radiology. A few years later Wilsey[84] improved upon the Potter-Bucky grid and described many of its physical characteristics which are valid today. The history of the development of the grid was published with first-hand authorization by Potter in 1931.[85]

4.4 Additional Equipment Improvements

Coolidge's tube, Snook's device, and the Potter-Bucky grid were the major developments of the first two decades of Roentgenology. There are a number of other developments that should be mentioned. The fluoroscope was developed by Thomas A. Edison, and according to his assistant, Mr. William H. Meadowcraft,[18] it was named by him. Edison is also responsible for the development and introduction of tungstate of calcium, a more highly fluorescent material than platinobarium cyanide. Pupin[16] lays claim to the first application of an intensifying screen-film combination. This was in 1896, but it was not until many years later that this technique was widely adopted. Charles L. Leonard found that by exposing two glass x-ray plates with the emulsion surfaces together, sealed in a light tight envelope, that the image was considerably better than that available with the single emulsion surface. This, according to Hickey,[18] was in 1904, but the double emulsion film was not available commercially until 1918.[63] Thereafter, double-coated film used with two intensifying screens became the standard imaging arrangement. The conversion from glass plates to film occurred during World War I when the glass from which the plates were made became scarce. Initially, radiographic films had a cellulose nitrate base. This composition was highly flammable and contributed to numerous hospital fires. It was replaced in the 1920's by cellulose acetate which today has largely been replaced by polyester base film.

5. CONTINUING REPORTS OF RADIATION INJURY

By the end of the decade of the 1930's, one would have expected that concern for radiation protection would have diminished. Numerical protective standards had been established and were generally accepted. Improvements in radiographic equipment and in protective design features had considerably reduced exposures to operational personnel. The classical reports of acute radiation injury had all but disappeared from the literature.

With the entrance of the world into the atomic era in December of 1940, some scientists recognized the need for more awareness of possible long-term radiation effects. During World War II there were reports from armament factories of accidental over-exposure of personnel who were untrained and unaware of the hazards of x-radiation. These reports were isolated, however, and were not at all like the disastrous effects reported on the Roentgen pioneers, on the miners in the Joachimsthal mines (the mines of Czechoslovakia that supplied the world's first radium), and on the girls employed in New Jersey and Pennsylvania to paint watch dials with luminous radium paint. The atomic age, however, offered the prospect of exposing large numbers of persons to ionizing radiation; therefore, the absence of acute effects did not mean, to some at any rate, that more refined safeguards were not necessary. The safeguards to be considered, for the most part, were directed to larger and larger groups of persons exposed to ionizing radiation. The concern turned slowly from that for the individual to that for society and, in fact, that is the major thrust of radiation control considerations in diagnostic Roentgenology today. Stone[27] showed a great perception when he said:

"...for the individual there is very little cause for worry, but for the future of the human race, with an ever increasing application of radiations, there is a real problem."

Stone was referring to Muller's[86] earlier statement:

"We must remember that the thread of germ plasm which now exists must suffice to furnish the seeds of the human race even for the most remote future. We are the present custodians of this all-important material, and it is up to us to guard it carefully and not contaminate it for the sake of an ephemeral benefit to our own generation."

Scientific papers dealing with radiation control did not cease with the atomic age. The frequency of these papers has not decreased to the present time and, in fact, may be increasing. Since the 1931 recommendations of the NCRP,[70] the generally accepted levels of radiation exposure for occupationally exposed persons has been reduced three times to a level of one tenth of those recommendations. It is quite possible that the current controversy over excessive medical exposure to patients and inadequate safeguards for industrial radiation applications may result in a still further reduction of the maximum permissible exposure. Present day concern is directed towards protection of the population at large and not specifically of occupationally exposed persons. The continuing provocation for radiation control in all situations is due to continuing reports in the scientific literature of radiation injury. The injuries for the most part are not acute but are subtle, long-term, hard-to-measure radiation effects. Refined epidemiological and statistical techniques are required for proper analysis of many of these effects.

Aside from the handful of serious radiation accidents associated with the atomic age, these continuing reports are generally associated with hematologic depression, leukemogenesis, carcinogenesis, cataractogenesis, infertility, life-span shortening, and the effects of radiation on the unborn fetus. Reports of acute skin erythema, desquamation, and epilation have been essentially absent from the medical literature for 30 years.

5.1 Hematological Effects

In 1906, a scant decade from the discovery of x-radiation, Warthin[87] published a very thorough review on the effects of x-rays on the blood forming organs. His review was prompted by over 60 reports on the application of x-rays for the treatment of leukemia, and he reviewed an additional 60 reports dealing with the effects of x-rays on the hematologic system, including some experimental evidence of his own. He concluded that "the therapeutic use of Roentgen rays in leukemia seems of doubtful value, or even dangerous." The deleterious effects of x-radiation on the hematologic system was therefore identified very early in the history of this new medical specialty.

Hardly a month went by during the first few decades of Roentgenology without a report appearing in the literature relating some blood disorder in exposed persons. The effects of high dose acute exposures to Roentgen rays on the hematologic system were well documented. The documentation was generally based on experiences with x-ray therapy of humans,[62,88-90] and experiments employing laboratory animals.[91-102] The information presented from the experience with radiotherapy patients was generally consistent. The same cannot be said for some of the animal experiments. For example, in a 1921 paper by Leitsch[103] the author concludes,

"...that the diminution of lymphocytes which usually occurs when rats are exposed to the X-rays for a short time, is not due in any way to the influence of radiation, but is a fright reaction caused mostly by the unaccustomed noises of the X-ray room, and partly by the manipulations necessary in obtaining material for the blood count."

Russ[104] took exception to these statements and conducted similar experiments with rats which allowed him to conclude, "...that the circulating lymphocyte (not only of the rat, but of many of the smaller animals) is a very vulnerable type of cell to X-rays..."

Perhaps the first reference to hematologic depression in radiation workers was presented by Portis[105] in 1915. Many later reports[45,106-113] were in general agreement that persons employed in medical radiation facilities were subject to chronic leukopenia. Mottram and Clarke[106] quantitatively described leukopenia in 20 radiation workers and showed that it was radiation induced since it commenced after only a few weeks of employment. They further noted that the leukocyte concentration increased and approached its pre-employment levels following extended vacation periods. It is somewhat surprising that these chronic effects on health were not adequately appreciated by the radiation scientists of the 1920's. All that Amundsen[111] could say was, "Improved means of protection are recommended so as to lessen the risk always incurred by all radiological workers."

Mottram[108] suggested in 1921 that routine blood examinations be employed in order to decide "whether or not the worker is being subject to over-exposure, or, alternatively, whether or not the devices designed for his protection are sufficient." Larkins[114] in 1921 reported a case of acute aplastic anemia in a radiation worker. This particular report was of interest because of the note of caution contained in the closing commentary:

> "This radiographer had recently changed over from soft tubes to the hard tubes for the greater part of his work, so the warning to be got from this case is that the protection hitherto employed in the use of harder tubes is not efficient. Further, it would appear advisable for radiographers and all those who are employed in radiographic work, or with radium, to have their blood examined periodically, say, every six months. As by so doing, the disease could be recognized at a very early stage and probably be stopped."

Interestingly, the recommendation for routine blood examination was adopted by the NCRP[70] and was not abandoned until the recognition by Loutit[115] and others in the middle 1950's that a dose of 25 rads or greater was necessary to produce significant hematologic depression.

The abandonment of routine hematologic examinations of radiation workers was also prompted by the enormous pile of records of blood counts of radiation workers on the Plutonium Project.[116] As Dr. Stone, Director of the Health Division of that project, stated in 1952,[27]

> "My medical colleagues and I decided to make such counts mandatory, and as a result, thousands of records of blood counts are now available, but the chances of finding anything of significance from them seems so remote that no one wants to spend much time analyzing them."

Jacobson and Marks[117] did study representative samples of these records and reported no hematologic changes in Plutonium Project personnel who received exposures not exceeding the "tolerance range", i.e., 0.1 R/day. However, they did report interesting results of experiments with mice, rabbits, and guinea pigs. No hematologic response was observed in these animals over a three-year period during which they received exposures of 0.1 R/day. At a daily exposure of 1.1 R, a minor but insignificant hematologic depression was seen. At 2.2, 4.4, or 8.8 R/day significant hematologic changes occurred. These findings led these authors to conclude:

> "Reliance on studies of the hematological constituents of peripheral blood is dangerous, since no findings can be expected with exposure in this range. The appearance of hematologic changes in personnel working in the field of nuclear physics or medicine, which are referable to an acute or chronic exposure to radiations, should be interpreted as serious and probably should preclude further actual or potential exposure of the worker thus affected."

Moshman[118] studied the records of occupationally exposed personnel at the Oak Ridge National Laboratory, the Los Alamos Scientific Laboratory, and the Hanford Works. No significant relationships between occupational exposure and hematologic response were found.

Researchers[119,120] at the Massachusetts General Hospital investigated the use of film badge monitoring during the period 1948-1950 to determine its suitability in a radiology department. They found that film badge monitoring provided quantitative and qualitative information on radiation exposures to personnel, but they could establish no relationship between the recorded exposures (which ranged to 31 mR/wk) and leukocytic depression.

On the basis of this accumulating evidence or, rather, lack of evidence, the NCRP in 1954[121] recommended that the practice of routine blood examination of occupationally exposed persons be discontinued.

All reports during this period were not negative. Nuttall[122] reported on his long-term experience with routine blood counts at the Manchester Radium Institute. From 1921 to 1932 the Institute was housed in grossly inadequate quarters. The rooms were small, the lighting and ventilation were very poor, and the layout was such that it was difficult for radium to be kept at an adequate distance from operating personnel. The average white blood cell (WBC) count of the workers in the Institute had, over that period of time, dropped to 4300, with individual counts as low as 3000. In 1933, when their new, modern, and carefully planned quarters became available, the average WBC rose to 5500 within six months. Over the next several years, Nuttall was able to trace the average WBC count and correlate its variations with changes in workload by members of his staff. On occasion he granted vacations of a month in duration for members of his staff in order to allow their WBC level to return to normal. Nuttall considered a WBC count of 6000 to be the "demarcation line between danger and safety." These experiences were reported for radium handlers. As Nuttall stated,

> "Workers in the X-ray therapy department using completely protective tubes, operative at 240 kV, and in the diagnostic department, have shown no particular blood changes, as a result of their employment."

Mayneord[123] reported that a significant leukocyte depression occurred in the peripheral blood of workers whose average weekly exposure exceeded in the range of 50-125 mR/week there was a slight, but probably not significant, leukocyte depression below normal levels. Paterson[124] also reported on and discussed the problem of hematologic responses to occupational levels of radiation exposure. Positive responses were reported at facilities where protective controls were loose and where high exposure type jobs existed.

A recent review by Wald et al.[125] has described the dose related hematologic effects in man. Their data were drawn from approximately 30 radiation workers acutely exposed to high doses as well as others exposed to low dose radiation. By categorizing the hematologic responses into five exclusive groups, they show how early hematologic responses to radiation may be helpful prognostic indicators of latent effects.

Currently, reliance on biological monitoring methods has been replaced almost completely by physical methods of personnel monitoring because of the inability to detect chronic, low-dose x-ray exposures by biological methods such as peripheral blood counts. In the late 1950's cytogenetic techniques were developed that have proven to be more sensitive indicators of radiation exposure than gross hematological findings. Some researchers are currently advocating the use of cytogenetic analysis as a biological radiation dosimeter.[126-130] If, in fact, this approach is adopted, it will no doubt have to wait for computerized techniques of chromosome scanning and analysis.[131,132]

5.2 Leukemogenesis

It is generally accepted that radiation exposure can cause leukemia.[133,134] This has been demonstrated with laboratory animals as well as with human population groups. The anticipation of increased leukemogenesis following low doses of radiation is a major contributor to the current controversy surrounding present radiation protection standards. As with other latent effects of radiation, the level of effect is often expressed as an incidence rate, i.e., cases of leukemia per population size per unit time per unit dose. The effect is also frequently identified as a relative risk value which is determined by computing the ratio of the frequency of leukemia in an irradiated population to a non-irradiated population. Use of an incidence rate requires the assumption of a linear dose-response relationship, which for most latent effects, including leukemia, is acceptable. Use of a relative risk value indicates that insufficient dose information is available; therefore, all that is possible is a direct comparison between the irradiated and the unirradiated population. A relative risk of 1.5, for example, would indicate a 50% higher incidence in the irradiated population than in the control population.

Although in 1942 Dunlap[135] published a rather extensive review on the effects of radiation on hemopoietic tissue and included a report of 24 cases of human leukemia, the first epidemiological study of radiation and leukemia was published in 1944 by Henshaw et al.[136] The death records for physicians, as published in the *Journal of the American Medical Association*, were compared with similar data for the general population obtained from the United States Bureau of the Census. Physicians were assumed to be the group at risk since many used x-rays and radium in their routine work. Analysis of the data covering a period of 10 years, 1933 to 1942, showed that 0.53% of the physician deaths were due to leukemia, while only 0.39% of the deaths in the general population were due to leukemia. These data resulted from nearly 27,000 physician death certificates and over one million deaths in the general population. When the data were adjusted for age specific rates, the relative risk factor for leukemia in physicians was 1.7; that is, leukemia occurred in the physician population 1.7 times more frequently than in the general population. These authors were quick to point out, however, "If we accept the available evidence that leukemia is higher among physicians than among the general population, it does not follow necessarily that radiation acted as the inciting agent." The findings of Henshaw et al.[136] were soon corroborated by Ulrich[137] and by Dublin and Spieglemann.[138]

In a letter to the editor of the *Journal of the American Medical Association*, March[139] pointed out that the primary reason for the twofold increase in leukemia among physicians as previously reported in 1944[140] was the inclusion of radiological physicians. When the radiologists were excluded from the physician study population, the leukemia incidence was approximately the same as the general population. In 1950, March[141] published the results of an expanded study. He analyzed the death notices in the *Journal of the American Medical Association* and identified the risk group as those physicians who were members of radiological societies. The control group consisted of all other physicians, presumably non-radiologists. During the two decades spanning 1929 through 1948, 334 non-radiological physicians died of leukemia, while 14 radiologists had a similar cause of death. These leukemic deaths resulted in an incidence of 1.51% for non-radiological physicians and 4.68% for radiologists, a nearly tenfold increase. This suggested a relative risk factor of approximately 10. More recent studies[142-144] have confirmed the findings of excess leukemia in radiologists.

Another population group providing convincing data on radiation-induced leukemia in humans are the atomic bomb survivors of Hiroshima and Nagasaki. Unlike the radiologists who, for the most part, received low dose intermittent radiation exposure, the survivors at Hiroshima and Nagasaki were exposed to rather high doses of radiation over a very short period of time. Several reports[145-154] offered by members of the Atomic Bomb Casualty Commission have analyzed this excessive leukemia incidence. Bizzozero et al.[154] have reviewed these data with reference to the incidence of leukemia during the initial 18 years post-exposure. There was a total of 326 cases of leukemia in survivors who were located within 10,000 meters of the hypocenter and 166 cases located within 1500 meters. Their analysis suggests that the higher the radiation exposure, as determined by distance from the hypocenter, the shorter the latent period for the development of leukemia. Acute lymphocytic leukemia was the most prominent cause of death. These authors showed that the annual leukemia incidence varied from 1.24 to 48.12 cases per 100,000 population, depending upon radiation dose and time interval analyzed. They did not estimate a numerical relative risk factor. More recent analyses[153] extending the analysis through 20 years post-irradiation have not significantly altered these findings.

Authors of studies of leukemia in atomic bomb survivors generally have been reluctant to postulate dose-response relationships as linear, at least down to a dose of 100 rads. The 1964 UNESCO report,[155] summarizing the then available data, suggested a linear relationship between dose and effect at least down to a level of 100 rads. The estimated average rate of increase in the incidence of leukemia was 1.1 cases/10^6 persons/year/rad at Hiroshima and 1.6 cases/10^6 persons/year/rad at Nagasaki.

Leukemia in patients receiving large field, high dose x-ray therapy for ankylosing spondylitis was the subject of a rather extensive special study by Court-Brown and Doll.[156] The study population included 11,287 men who were treated in this fashion between 1935 and 1954. The radiation exposures ranged from 112 R to more than 3000 R administered over a fractionated course of treatment. On the basis of 37 cases of leukemia a regression line with a slope of 0.5 cases/10^6 persons/year/rad was calculated. The radiation dose used in this analysis was the estimated mean bone marrow dose.

Through the early 1950's the condition of thymic hypertrophy in children was often treated therapeutically with local x-irradiation. The number of cases of leukemia reported[157] in surveys of these children was small; therefore, definitive estimates of risk were not possible. The suggestion was strong, however, that these patients did suffer an increased level of leukemia over unirradiated controls.

Other sources of radiation exposure have been implicated in excess leukemia in humans. In their 1962 review, Wald et al.[125] reported an apparent excess leukemia in patients treated for polycythemia vera with P-32 and hyperthyroid patients treated with 1-131. Pochin[158] in a broader study of patients treated for thyrotoxicosis with 1-131 also described increased leukemia in these patients. Other studies[159-161] contain inferential evidence of excess leukemia in irradiated persons, but in each case, the population size was not large enough to allow significant statistical analysis.

5.3 Leukemia Following Diagnostic Roentgen Examination

The identification of elevated levels of leukemia following relatively high doses of radiation is rather certain and has been shown conclusively in the several human population groups previously discussed. As Cronkite et al.[162] concluded in 1960:

> "Data are now adequate to indicate that, for the high level, single dose exposure of man, the incidence of leukemia is approximately linear with dose. At dose levels of perhaps 100 R equivalent or greater, the incidence is approximately 1 to $2/10^6$ persons at risk/year/rad, at least from approximately the 2nd to the 15th year following exposure. Below dose levels of approximately 100 R equivalent, the available data are inadequate for prediction."

Indeed, although the evidence of radiation-induced leukemia is unequivocal at dose levels above 100 rads, there is little direct evidence to support a similar relationship following a few rads or even tens of rads. However, considerable concern over the latent effects of diagnostic levels of radiation have been expressed since evidence is mounting that even at the low levels of radiation exposure employed in diagnostic Roentgenology, elevated levels of leukemia follow.

Epidemiological analyses indicate that an increased risk of developing leukemia exists following diagnostic Roentgenographic examinations of adults. For the most part, these have been retrospective studies. Stewart el al.[163] investigated the radiation history of nearly 1000 cases of leukemia in adults. They concluded that 8% of the cases other than lymphatic leukemia were radiation-induced and exhibited an average latent period of 3 to 4 years. Diagnostic x-ray exposures of the chest and abdomen were implicated as the possible causative factors. Stewart et al.[164] also implicated postnatal diagnostic x-ray in observed excess leukemia deaths at the age of two and three years. Neumann[165] reported a relative risk for leukemia of 1.7 in tuberculosis patients who presumably were exposed for chest films more frequently than the control population. His risk estimate, however, was based on only ten cases of leukemia over a six-year period.

Polhemus and Koch[166] studied the medical records of 251 leukemic children seen in their clinic between 1950 and 1957. Prior therapeutic or diagnostic irradiation was significantly more frequent among the leukemic children than matched controls. On the other hand, Birch and Baker[167] observed no excess leukemia in 1480 children who received repeated fluoroscopic examinations. Similarly, Murray et al.[168] observed no excess leukemia in children irradiated postnatally with diagnostic x-rays; however, excess leukemia was evident in therapeutically irradiated children. Their study consisted of over 6400 children.

In a more recent epidemiological study[169] attempting to relate diagnostic radiation exposure with the subsequent development of leukemia, relative risk values based on nearly 200 cases of leukemia varied between 1.3 and 3.8, depending upon the anatomical site of examination. An attempt to demonstrate increasing relative risk with increasing number of films per examination failed. Descriptive dose-response relationships were not possible in this study because of the small number of cases and because of the lack of radiation dosimetry.

5.4 The Effects of Radiation on the Fetus

There is mounting evidence that diagnostic levels of radiation produce harmful effects in the irradiated fetus that may not be manifested until years later. Specifically, the major effects of concern are genetic damage and leukemia induction. The reviews of Russell and Russell,[170] Yamazaka,[171] Rugh,[172] and Van Cleave[173] have described the effects of high-dose radiation exposure on the fetus during the various stages of pregnancy. Collectively, these reviews show that the fetus is most sensitive to ionizing radiation during the first trimester and exhibits maximum radiosensitivity immediately postconception. High-dose irradiation during the second trimester results in an increased incidence of congenital abnormalities and neonatal death. During the third trimester the fetus is most radioresistant, but exposure in this time interval may result in latent radiation injury, such as leukemia, cancer, and long-term genetic change.

The effects following high-dose irradiation are well described and generally accepted. The effects of low-dose irradiation are in doubt, because the available data are not conclusive. Nevertheless, this doubt has stimulated a major controversy among radiation scientists about the effects of these low doses of radiation on the fetus and what, if any, additional radiation protection measures should be instituted.

Sternglass,[174-178] Morgan,[11,179,180] Gofman,[181] and Tamplin[182] are outspoken proponents of the potentially disastrous effects on the population due to fetal and childhood irradiation with very low levels of exposure. The effects suspected by these scientists are difficult, if not impossible, to measure directly. However, on the assumption of a linear non-threshold dose-response relationship for morbidity and mortality, estimates can be made of the extent of these harmful effects. As our society expands technologically, it is expected that the population will be subjected to increasingly higher radiation exposures with the greatest harmful effect resulting from *in utero* irradiation. In order to ensure that these suspected effects do not, in fact, occur, say these authors, more careful and stringent regulation of medical x-ray exposure is required and a reduction of the maximum permissible dose for the population at large is called for.

Sternglass[174-178] has attempted to show that neonatal and infant mortality in the United States has been adversely affected by both diagnostic x-ray exposure and long-term exposures from nuclear weapons' fallout. He reports increased rates of childhood leukemia and a change in the age distribution at death in population groups exposed to low-level radiation. He also reports relatively higher rates of stillbirths, neonatal deaths, and infant deaths following radiation exposures reported to be less than one Roentgen. The hypothesis proposed by Gofman[181] and Tamplin[182] is that perhaps 16,000 additional cases of cancer are likely to occur each year if the population were exposed to currently accepted maximum permissible dose levels (170 mrem per year). They identify children, irradiation *in utero*, or in early life, as the portion of our population at highest risk. Their interpretation of previously reported data on these low-dose effects has been effective in discouraging continued development of nuclear power in this country and may in the future be effective in establishing stringent controls over medical radiation applications. Morgan[11] estimates that over 90% of the genetically significant dose in the United States from man-made sources of radiation comes from medical exposure. He further suggests[179,180] that between 3500 and 29,000 deaths per year may occur in the United States as a consequence of radiation injuries following diagnostic medical radiation exposure. He recognizes that medical x-ray utilization is essential and that many lives are saved by its use. However, he also estimates that the same benefits could be derived with perhaps one tenth the patient exposure if proper education, training, and certification of all physicians and x-ray technicians were required, along with proper equipment design and maintenance.

Although many of these statements have been questioned on a scientific basis, legislation is currently before the United States Congress to accomplish some of these suggestions. The Sternglass hypothesis is not widely accepted and has, in fact, been subject to rather convincing rebuttal.[183] Likewise, Bond[184] has recently refuted the interpretations of Gofman and Tamplin in a point-by-point fashion.

Although the pronouncements of some of these scientists have been refuted, and the estimates of others may be gross exaggerations, there is, in fact, a wealth of epidemiological data to show that diagnostic levels of x-radiation to the fetus do result in an increased risk of latent injury. By 1960[162] the total number of "adequately documented" cases of radiation-induced leukemia was reported to be 226, but with the introduction of epidemiological techniques, this level of response may be underestimated.

Perhaps the most authoritative investigation of cancer and leukemia in children following irradiation *in utero* is the Oxford Survey of Childhood Malignancy[164,185-189] under the direction of Alice Stewart. This survey is a continuing retrospective epidemiological study of all children born in England, Scotland, and Wales who died of leukemia or cancer before their tenth birthday. The study covers the period from 1943 to the present with the latest report[189] containing data through 1965. Each reported case of leukemia was matched with a healthy control on the basis of a number of geographic and demographic characteristics. Mothers of each case or control child were interviewed and hospital records were reviewed to determine the frequency and the extent of *in utero* irradiation. The latest estimates of radiation-induced malignancy are based on 7649 cases and 7649 controls. Diagnostic x-radiation *in utero* was administered to 1141 case children and 774 control children. These frequencies resulted in an estimate of relative risk of subsequent malignancy following *in utero* irradiation of 1.48. Stewart and Hewitt[186] showed that radiation-induced leukemia was identifiable on an epidemiological basis because of differences in age distribution at the time of diagnosis. The modal incidence of leukemia in unirradiated children was approximately two to four years, while that for irradiated children was four to six years. Stewart and Kneale[187] showed that the risk of malignancy following obstetric radiography remained relatively constant in a population of 12,694 children who were born between 1943 and 1965. Apparently, the development of better radiographic equipment and techniques during that time was offset by increasing use of obstetric radiography.

There are other studies similar to the Oxford Survey, and not all have resulted in positive correlations. Ford et al.[190] described increased malignant diseases in children before the age of ten following irradiation *in utero*. The study population consisted of 78 cases of leukemia and 74 cases of other malignant diseases. The cases and 306 matched controls resided in Louisiana during the years 1951 through 1955. *In utero* x-ray exposure had occurred in 26.9% of the leukemia cases, 28.4% of the cases of other malignant disease, and in only 18.3% of the controlled cases. Court-Brown et al.[191] confined their investigation to children dying only of leukemia following irradiation *in utero* and their approach was somewhat different. They contacted all women known to have received diagnostic x-ray examination of the abdomen or pelvis during pregnancy between the years 1945 and 1956. Children born to these women were then followed through 1958 and an identification made of those who died of leukemia. A total of 39,166 children were identified, and 9 of these were discovered to have died of leukemia within the specified time period. Based on national British leukemia rates, the expected number was estimated to be 10.5. Consequently, this study does not support a radioleukemogenic cause-effect relationship.

In a rather extensive epidemiological study conducted in this country, MacMahon[192] reviewed 734,243 birth records in 37 large maternity hospitals in the northeastern United States during 1947 through 1954. Approximately 11% of these children had received diagnostic x-irradiation *in utero*, and it was estimated that "cancer mortality was about 40% higher in the x-rayed than in the un-x-rayed members of the study population." Shortly after this finding, MacMahon and Hutchison[193] surveyed the then available ten studies involving irradiation *in utero* and latent malignant disease. As they point out, there were many differences among the ten studies, but relative risk estimates were gained from each. Five of the studies supported the hypothesis that diagnostic x-irradiation *in utero* results in elevated frequencies of malignant disease, while five of the studies did not support the hypothesis. As MacMahon and Hutchison point out, however, the five studies that resulted in relative risks less than 1.0 involved very small samples in which the sampling variability was large. They concluded that based on all of these studies, there was in fact an increased risk in children irradiated *in utero* of approximately 40% higher than unirradiated children. Table 4 contains a summary of the MacMahon-Hutchison review. Identification of the ten studies and the relative risks from each are included as reported by MacMahon and Hutchison except for the value from the Oxford Survey. The most recent estimate from that study is 1.48 instead of the previously reported 1.65.

Isolated case reports[199] continue to appear describing childhood leukemia following *in utero* exposure. However, results of studies appearing since the MacMahon-Hutchison review continue to be divided in their conclusions although, as before, the larger the study population, generally, the more positive the correlation. Jablon and Kato[200] reviewed the childhood malignancy experience of 1292 children exposed *in utero* in 1945 to the atomic bombs of Hiroshima and Nagasaki. They reported no excessive mortality from leukemia or other cancers. Ager et al.[201] detected no significant increased risk in the development of leukemia for either exposure *in utero* or postnatal

TABLE 4

The Relative Risk of Developing Latent Malignant Disease Following X-Irradiation *In Utero*- A Summary of Ten Studies

Investigation	Ref.	Relative Risk
Polhemus and Koch	166	1.33
Stewart et al.	188	1.48
Ford et al.	190	1.71
Court-Brown et al.	191	0.85
MacMahon	192	1.42
Kaplan	194	1.72
Kjildsberg	195	0.59
Lewis	196	0.42
Murray et al.	197	0.92
Wells and Steer	198	0.72

From MacMahon and Hutchison.[193]

exposure. Their analysis was based on 112 cases of childhood leukemia.

On the other hand, the study of Graham et al.[169] resulted in significant risk differences for not only *in utero* and postnatal irradiation but also for preconception irradiation. The study population consisted of all reported cases of leukemia in children up to age 15 during the years 1959 through 1962 from the urban population centers of New York state exclusive of New York City, and from the Baltimore and Minneapolis-St. Paul metropolitan areas. Case identification was made on the basis of examination of the medical records and by interview of the mothers of the children with leukemia. The conclusions of the study were based on a total of 319 cases and 884 controls which included only individuals for whom definitive information for all aspects of the study were available. Relative risk values of developing leukemia following diagnostic levels of x-irradiation were determined for a number of conditions and are summarized in Table 5. In addition to the relative risk values reported in Table 5, Graham et al.[169] showed that their analysis suggested an even higher relative risk for combinations of radiation exposure. Preconception irradiation plus *in utero* irradiation carried a higher relative risk than either alone. Higher relative risk values were observed following irradiation in the third trimester, rather than the first trimester. Following postnatal irradiation to one site, the excess leukemia was approximately one half that when x-irradiation of two or more sites was conducted.

5.5 Summary of Radiation-Induced Leukemia

On the basis of the evidence generated by studies of the types just discussed, it can be concluded that high doses of radiation, for example, greater than 100 rads, can cause leukemia. The effects of low doses, such as those employed in diagnosis, are unknown although the available evidence suggests that no safe threshold dose exists. The primary difficulty in obtaining conclusive data rests on limitations in the size of the study populations. Table 6 is from Buck's[202] discussion of the minimum population size necessary to detect a significant effect at the dose levels indicated. It should be noted that if a population received a 5 rad dose, the study would require that 6,000,000 individuals followed for one year, or 600,000 persons followed for ten years before statistically positive results could be obtained. These values assume a linear non-threshold dose-response relationship with a slope of one case/10^6 persons at risk/rad/year. The ICRP has concluded:[203]

> "On the assumption of a linear relationship, the total leukemia risk would appear to be of the order of 20 cases per million persons per rad. Longer periods of observation may suggest that this is an underestimate for high dose rates. However, it may be an over-estimate for low dose rates."

Auxier et al.[204] have reported revised dosimetric data for the atomic bomb survivors that suggest a leukemia level of 50 cases per 1,000,000 persons per rad might be more accurate.

On the other hand, Marinelli[205] cites some experimental evidence in laboratory animals suggesting a lower leukemogenic effect at low dose rates. Wise's analysis[206] "throws doubt on the

TABLE 5

The Relative Risk of Developing Leukemia Following Diagnostic X-Irradiation Under Various Conditions

Irradiation Condition	Relative Risk
No radiation	1.00
Preconception irradiation of mother	1.55
Preconception irradiation of father	1.31
In Utero irradiation (all sites)	1.40
X-ray pelvimetry	2.00
Postnatal irradiation (all sites)	1.19

From Graham et al.[169]

TABLE 6

Minimum Population Sizes for Various Specified Doses of Radiation Necessary to Detect Significant Increases in Leukemia- The Calulation Assumes a Linear, Non-Threshold Dose Response Relationship With a Slope of 1.0 case/10^6/rad/ year.

Dose from birth to age 34 (rads)	Minimum Person – years at ages 35 to 44
5	6,000,000
10	1,600,000
15	750,000
20	500,000
50	100,000
100	30,000
200	10,000

From Buck.[202]

assumed linear dose-response relationships" because of the postirradiation leukemia incidence distribution in time.

Gunt and Atkinson[207] reviewed the literature on radiation-induced leukemia in 1964 for the purpose of an editorial in *Radiology*. Their conclusions were stated as follows:

"It would appear in summary of the present position that there is insufficient evidence to assess the precise extent of the leukemia danger from medical radiation; nor is it likely that the necessary data will be soon forthcoming. Nevertheless, there is no room for complacency, for almost certainly some radiation-induced leukemias would have been avoidable in the past had greater care been taken in preventing unnecessary or excessively large exposures. The need is for continued vigilance whenever medical radiation is employed."

A more recent editorial[208] calling for prudence in the application of diagnostic x-rays concluded:

"The need for some demonstration of concern and selfpolicing within the medical community seems clear. It is incumbent upon medical radiologists to impose controls and to provide education directed toward protection of the patient. If they do not, others will do so."

5.6 Carcinogenesis

The previous discussion relating radiation and leukemia holds equally well for other radiation-induced malignant diseases. The two major study groups that have produced data to support a radiation-cancer relationship are the atomic bomb survivors and patients irradiated therapeutically for various conditions of the head and neck.

For a three-year period ending in 1961, Socolow et al.[209] reviewed the results of routine medical examinations on over 15,000 survivors in Hiroshima and Nagasaki. Enlarged thyroid glands were reported in 355 patients, and selective biopsies disclosed 21 cases of thyroid carcinoma. Because of the difficulty of establishing a control population in this instance, definitive conclusions were not possible. Another study[210] of atomic bomb survivors reported an investigation of 1253 thyroid specimens from adult autopsies and 342 surgical pathological thyroid specimens for the period 1948 through 1960. Again, due to difficulties in experimental design, precise conclusions were not available. There was, however, an excess of thyroid carcinoma in specimens from patients exposed within 1400 meters of the hypocenter compared to the unexposed controls. Wanebo et al.[211] have reported that breast cancer among the Japanese atomic bomb survivors increased by two to four times over controls when the radiation dose received exceeded 90 rads. These authors reported similar findings relative to radiation-induced lung cancer.[212] In a more recent review of the findings of the Atomic Bomb Casualty Commission (ABCC)[213] Miller states:

"One may conclude that, among the Japanese survivors of the atomic bomb, only leukemia and thyroid cancer have been shown to be radiation-induced. The evidence pertaining to cancer of the breast or lung is still very much in doubt."

Other investigators have reported increased neoplasms in irradiated patients; however, owing to the small populations available, the significance of the reported risk factors is not great. Saenger et al.[214] reported a relative risk of 91 for the development of thyroid carcinoma from a series of 1644 patients who received radiation therapy for benign conditions of the head, neck, and chest. An analysis[215] of the data of Takahashi et al.[216] resulted in relative risk factors ranging to 19.9 for thyroid cancer, to 4.5 for other cancers of the neck, and to 27.4 for skin cancer in patients receiving therapeutic irradiation with external beams. Generally, for each category of response, the value of relative risk increased with increasing radiation dose. Brinkley and Haybittle[217] found an increased incidence in cancer at the site of irradiation in 277 patients treated by an artificial x-ray menopause during 1942-52. Raventos and Winship[218] reported an average latent period for the development of radiation-induced thyroid cancer of 10.9 years in a study of 528 cases. Sagerman et al.[219] were unable to identify a consistent latent period in 21 cases of radiation-induced neoplasia in children treated for retinoblastoma with radiation therapy. These 21 cases developed in 397 patients who were so treated.

Hemplemann[220] has provided some rather striking data that show an increase in thyroid carcinoma with increasing dose when radiation is locally administered to the thyroid region. Three study populations were utilized. The low risk group included 786 children who received one or more radiation therapy treatments for thymic enlargement. The thyroid dose was of the order of 20 rads and eight thyroid tumors were detected in this group of patients. The second population group included nearly 3000 persons treated for thymic enlargement in one radiation therapy facility that employed a treatment modality which resulted in an approximate thyroid dose of 335 rads. Increased nodularity and carcinoma were demonstrated in this population. The third group included 19 children who lived on Rongelap Island in the South Pacific and were exposed to fallout from one of the nuclear tests at Eniwetok Island. The estimated dose from both external and internal radiations was in the range of 700 to 1400 rads. Within 12 years of exposure 15 of the 19 children had developed thyroid modules, but no carcinoma had developed. Hemplemann summarized his findings in the following way:

"Because of the limited number of cases of carcinoma in the three studies, we can say little about the dose response of radiation-induced malignant lesions. The risk of developing carcinoma in the three studies ranges from 0 to 5.5 cases and that for nodularity 38 to 52 cases per 10^6 persons exposed per rad per year."

5.7 Life Span Shortening

Because of the many radiation-induced malignancies and deaths in early radiological personnel, it was expected that this occupational group was experiencing a higher death rate than other persons. Unexpected was the finding that this higher death rate was not only due to malignant disease but also to non-specific premature aging. Reviews of animal experiments[221-223] have shown radiation to be an agent capable of producing premature aging and death. Some of the more recent animal data from which estimates of life span in man are made have been summarized by Grahn and Sacher[224] and are shown in Table 7. As seen in this table, the radiation-induced life span shortening for mice is generally in the range of 20 to 50 days per 100 rads. This is approximately 0.05% life shortening per rad in the mouse which, when related to man, is 13 days life shortening per rad.

In 1947 and 1948 Dublin and Spiegelman[138,233] published two papers dealing with the mortality of American physicians based on the records of the American Medical Association between 1938 and 1942. They concluded that from age 35 "the prospective lifetime for physicians is somewhat less than that for white persons in the general population, but the differences are not significant." The greatest category of excess mortality was that from leukemia in radiologists. The finding of increased leukemia in radiologists was also reported by Dunlap,[135] Ulrich,[137] March,[139-141] and Warren.[142,144] In 1956, Warren[142] also showed that there was generalized life span shortening in this population, not by any one specific disease, but apparently by premature non-specific aging. This finding was criticized by Seltser and Sartwell[234] for the statistical methods employed. They concluded:

"The fact that the average age of radiologists at death is younger than that of other physicians cannot be ascrib-

TABLE 7

Life Span Shortening Following Single Whole Body Exposures of Mice

Strain	Sex	Radiation	Life Span Shortening per 100 rads (days)	Investigator	Ref.
Several different strains	M	98 kVp X-rays	15	Gowen and Stadler, 1956	225
BAF_1	M	80,135,250 kVp X-rays	25	Grahn and Sacher, 1958	226
	F		40		
Several different strains	M	200 kVp X-rays	28	Grahn, 1960	227
	F		81		
LAF_1	M	mixed gamma rays	37	Upton et al., 1960	228
	F		47		
SAS/4	M & F	15 MeV X-rays	40	Lindop and Rotblat, 1961	229
RF	F	250 kVp X-rays	24	Storer, 1962	230
BDF_1	F	250 kVp X-rays	45	Storer, 1965	223
RF	F	^{60}Co gamma rays	29	Spalding et al., 1967	231
RF	F	^{60}Co gamma rays	33	Upton et al., 1967	232
LAF_1	M	^{60}Co gamma rays	18	Grahn and Sacher, 1968	224
	F		25		

From Grahn and Sacher.[224]

ed to their exposure to ionizing radiation, since differences in age composition alone can account for the finding."

Later reports by Warren[144,235] confirmed his earlier findings. In these reports he showed that the average age at death for radiologists during the period 1934-1939 was 56, while that for the U. S. adult white males during the same period was 62. During the 1940's this difference in age at death was reduced to approximately one year. Currently, according to Warren,[144] there is no significant difference,

"...probably as the result of better regard for safety standards... From these findings, one may conclude that current occupational maximal permissible dose levels provide adequate protection."

One difficulty with these types of studies is the selection of adequate control groups. Seltser and Sartwell[236] attempted to overcome this problem by identifying three groups of physicians, each at a different level of risk of radiation injury. The highest risk group consisted of members of the Radiological Society of North America–practitioners in the use of radiation equipment. Members of the American College of Physicians were identified as the intermediate risk group since only a small percentage of these physicians used radiation equipment. The lowest risk group was composed of members of the American Academy of Ophthalmology and Otolaryngology. These were presumably physicians who rarely, if ever, came into contact with radiation. The physician population making up these three risk groups totaled nearly 18,000 and consisted of all who became members of these societies from the time of their organization through 1954 and whose survivorship history through 1958 could be determined. The data analysis resulted in significant differences in mortality as a function of the medical specialty. Table 8 is extracted from their figures and shows that although the difference in mortality experience between the high risk and the low risk populations is decreasing, there existed a statistically significant difference through the time period 1955 to 1958. The authors stated:

"The force of the conclusion that X-ray exposure is responsible for the difference in mortality is greatly strengthened by the fact that before the study was undertaken the hypothesis to be tested had been stated. The observed gradient of mortality supports that hypothesis in detail."

The mortality gradient, however, is due to the mortality experience of older physicians. There was no difference in mortality among the three physician specialties during the 1955-1958 period for physicians less than 50 years of age.

Mortality studies[213,237-239] of the Japanese survivors of the atomic bombings have demonstrated increased mortality in the 99,393 persons of the exposed study populations. Even when leukemia deaths were excluded from the analysis of persons exposed within 1200 meters of the hypocenter, the mortality rate was increased by approximately 15%. Increased mortality was also shown to be related to distance from the hypocenter and, therefore, presumably related to radiation dose.

All studies designed to show a relationship between radiation exposure and increased mortality have not been positive. Among these is a report by Court-Brown and Doll[240] of their analysis of mortality in British radiologists. This study was prompted by Warren's[144] results showing increased mortality in American radiologists. The Court-Brown and Doll study population consisted of 1377 male British radiologists who died during the period 1897 to 1957. Their study and analysis showed:

"...no evidence that occupational exposure to ionizing radiations has caused a detectable non-specific shortening of the expectation of life. A significant excess of cancer deaths was found among those entering the practice of radiology before 1921."

TABLE 8

Mortality Experience of Members of Three Physician Specialties - the Radiological Society of North America, The American College of Physicians, and the American Academy of Ophthalmologists and Otolaryngologists

Calendar Period	RSNA	ACP	AAOO
	Age Adjusted Death Rates Per 1000 Person-Years		
1935-1944	18.4	15.4	13.0
1945-1954	16.4	13.7	11.9
1955-1958	13.6	11.4	10.6
	Median Age at Death		
1935-1944	71.4	73.4	76.2
1945-1954	72.0	74.8	76.0
1955-1958	73.5	76.0	76.4

From Seltser and Sartwell.[236]

Similarly, a study[241] of the mortality of New England dentists between the years 1921 to 1960 showed no relationship between mortality and exposure to ionizing radiation. Nearly 7000 males trained as radiological technologists by the U. S. Army during World War II have been followed in an attempt to detect latent radiation injuries.[242] All findings to date have been negative. Griem et al.[243] followed 1008 children who had been irradiated *in utero* during 1948 with a dose estimated to be between 1.5 and 3 rads. Their findings did not reveal increased mortality or decreased life span.

In summary, the general consensus is that radiation exposure shortens life. The precise dose-response relationship is unknown, but many investigators have suggested life shortening times per unit dose which implies a linear relationship. Jones[244] calculated that human life would be shortened by 15 days per Roentgen exposure. The calculation was based on available animal experimental data. Later, Failla and McClement[245] applied the Gompertz function to the experimental data of Lorenz et al.[246] which had been obtained from chronically irradiated mice. They arrived at a life shortening level of one day per Roentgen exposure. The Space Radiation Study Panel of the National Academy of Sciences - National Research Council[247] in their review of the available evidence suggest that for chronic low level exposure, human life may be shortened by ten days per rad.

These quantitative estimates of life span shortening in man are based primarily on experimental data from irradiated animals. Supportive evidence from mortality data from American radiologists confirms the principle of life span shortening but quantitative life shortening periods per unit dose are not possible from these data. Dose estimates for the accumulative exposure to early radiologists vary from quite low levels to 1000 rads. At the rate of one to ten days of life shortening per rad of radiation dose, it is possible that Warren's[144] finding of an average five-year depression in life span for radiologists dying in the 1930's was true.

5.8 Additional Human Radiation Responses

There are a number of additional human responses to ionizing radiation that have been reported. These responses include cataractogenesis, congenital malformation, fertility depression, and cytogenetic damage. The data available to establish that these human responses are radiation-induced are for the most part only suggestive and not conclusive. Nevertheless, they have been important considerations in our continuing revision of radiation protection standards.

The tragedy that befell many of the first cyclotron physicists was not adequately documented until the late 1940's.[248] These scientists suffered from disabling radiation-induced cataracts at an early age. The fault was the accepted practice at the time of remaining in the vicinity of the cyclotron beam, if not in the beam itself, in order to visually adjust the beam target apparatus. Cataracts have also been observed with high incidence in radiotherapy patients who received large doses to the lens of the eye.[249] Because of reports such as these, the lens of the eye is being given more consideration by advisory groups and more attention by radiologists and dentists during examinations of the head.

Increased incidence of congenital malformation in children born to radiologists and decreased fertility in radiologists were reported in two papers separated in time by nearly 30 years.[250,251] These observations were not subjected to tests of significance, however, and the sample size was relatively small.

Because of the implied genetic consequences of radiation, radiation-induced chromosome damage is of current interest. Many studies with cultured mammalian cells irradiated *in vitro* have resulted in a precise dose-response relationship based on cytogenetic damage. At doses exceeding 25 rads chromosome aberrations are readily visualized and have been reported in radiotherapy patients,[252-255] in radiation accident victims,[256] in the Marshall Islanders who were exposed to fallout,[257] and in the survivors of the atomic bombs of Hiroshima and Nagasaki.[258,259] Chromosome aberrations in human peripheral lymphocytes have also been measured following lower radiation doses in vitro[260-265] and in vivo following occupational levels of exposure.[266-268] Schmickel[265] has presented an informative review of cytogenetic technology including aberration formation and scoring techniques.

Chromosome aberrations have also been detected in peripheral lymphocytes of humans following diagnostic radiological procedures[269-273] and these observations are of particular concern to radiation scientists. There have also been negative findings of chromosome aberrations following diagnostic procedures,[274] but even the possibility of this type of damage must be considered. Chromosome aberrations are obviously undesirable and are probably deleterious at any level. Precise relationships between radiation dose and cytogenetic response have been made, but the relationship between cytogenetic response and any latent physiological damage is unknown. Until the latter relationship is identified, the level of hazard resulting from chromosome aberrations following diagnostic procedures will be unidentifiable.

VI. CURRENT RADIATION PROTECTION STANDARDS

The continuing reports of radiation injury and the current awareness of the late effects of low level radiation have resulted in a constant evaluation and review of our radiation protection standards. In general, this evaluation has resulted in not only a lowering of the level of radiation exposure acceptable for radiation workers but also in a more precise definition of qualifying conditions placed on the exposure. Several discussions[275-277] in the recent literature have attempted to define and identify various qualifying considerations of our current radiation protection standards.

In the United States our radiation control practices are promulgated primarily by the National Council on Radiation Protection and Measurements (NCRP), and Table 9 contains a chronological review of the whole body Maximum Permissible Dose Equivalent (MPD) from the time of the first recommendation, 1931, to the present. During the past 40 years the MPD has been revised downward three times to 1/10 of its original value.

In addition to the NCRP, there are two other rather important radiation advisory groups–the International Commission on Radiological Protection (ICRP) and the International Commission on Radiation Units and Measurements (ICRU). Some professional societies also occasionally publish radiation protection recommendations.

TABLE 9

Recommended Annual Whole Body Maximum Permissible Dose Equivalents for Workers Occupationally Exposed to Ionizing Radiation

Period	MPD (rem)
1931 - 1936	50
1936 - 1948	30
1948 - 1958	15
1958 - Present	5

6.1 Radiation Units

There are several units for expressing radiation quantity. The rad, an acronym for radiation absorbed dose, is the special unit of absorbed dose. It has a value of 100 ergs of energy absorbed per gram of absorbing material. There are no restrictions to the type of ionizing radiation, or to the type of absorbing material considered. The rad is the unit that is employed in radiobiological research since biological effects are generally related to the rad and it is also the unit used to specify radiation doses received by humans other than occupational exposure. Dose prescription for radiotherapy patients, dose estimations for diagnostic patients, and dose determinations employed for dose-response relationships in man are all generally expressed in units of rads.

The Roentgen (R) is the unit of exposure and applies only to the interaction between X or gamma radiation and air. The Roentgen is the oldest consistent unit of radiation measurement having its origin at the second International Congress of Radiology in 1928. It identifies the intensity of radiation by the number of ion pairs formed in air at standard temperature and pressure. It is applicable only for photon energies not exceeding 3 MeV and only for the interaction between those photons and air. This unit is most often employed in the calibration of radiation producing machines as a means of specifying their output intensity. Rigorously, this quantity has been defined as follows:[278]

> "The exposure (X) is the quotient of ΔQ by Δm where ΔQ is the sum of the electrical charges on all the ions of one sign produced in air, when all the electrons (negatrons and positrons), liberated by photons in a volume element of air, whose mass is Δm are completely stopped in air. The special unit of exposure is the Roentgen (R)"

$$1 \text{ R} = 2.58 \times 10^{-4} \text{ C/kg}$$

This value for the Roentgen is numerically equal with the earlier stated value[279] of 1 esu of charge per 0.001293 g of air at standard temperature and pressure. It is also equal to the often-quoted value of 1.610×10^{12} ion pairs per gram of air. The Roentgen is sometimes used to describe the exposure at the surface of a patient during diagnostic procedures. Usually, if the exposure is known, the dose (in rads) can be computed if beam quality, field size, and other factors are known. For example, an abdominal film at 85 kVp, 100 mAs may result in an exposure of 300 mR at the position of the entrance surface of the patient, but the resulting midline dose may be only 50 mrads and the gonadal dose only 5 mrads.

Exposure of radiation workers is expressed in units of rem, an acronym for radiation equivalent man. The recommended dose, or limiting dose, is identified as the maximum permissible dose equivalent (MPD) and it is expressed in rem. The rem is employed in order to be precise in the specification of occupational radiation exposure. It applies to all types of ionizing radiation and the manner in which they are presumed to interact with man. The use of the unit rem is necessary because some types of ionizing radiation to which workers are exposed produce more severe effects on man than do X and gamma rays on a per rad basis. For example, an absorbed dose of 1 rad from neutron radiation produced by a nuclear reactor may result in a human response equivalent to ten rads from X or gamma rays. Consequently, the dose of one rad resulting from this neutron exposure would carry a value of 10 rem. The multiplication factor 10 is known as the quality factor (QF) and was formerly identified as the relative biological effectiveness (RBE). The term RBE is now reserved exclusively for experimental radiobiology. Therefore, the relationship between the radiation absorbed dose (D) and the dose equivalent (DE) is:

$$DE = (D) \times (QF) \times \text{other factors}$$

The other factors are dose modifying factors, particularly those associated with the temporal and spatial distribution of radiation dose. For practical application of these recommendations in diagnostic radiographic procedures, the QF may be considered to have a value of 1.0. Similarly, unless there is evidence to the contrary, the other dose modifying factors may be considered to be 1.0. Consequently, in diagnostic radiology it is accepted that an exposure of 1 Roentgen will result in an absorbed dose of 1 rad which will then be equal to 1 rem.

Radiation quantities experienced in diagnostic radiology are usually lower than 1 R, 1 rad, or 1 rem; therefore, units 1/1000 of these values are used, i.e., milliR (mR), millirad (mrad), and millirem (mrem). A more complete discussion of these radiation units will be found elsewhere.[278]

6.2 National Council on Radiation Protection and Measurement

The NCRP was formed in 1931 as the United States Advisory Committee on X-ray and Radium Protection. This Committee was modeled after the British Committee of a similar title that had been established in 1921. The Committee structure required eight members. Dr. Lauriston S. Taylor was its first chairman and has continued in that capacity to the present day. He has also described the development of this organization.[280] There were two representatives from the American Roentgen Ray Society, two representatives from the Radiological Society of North America, and one representative from the American Medical Association. An additional two members were on the Committee representing x-ray equipment manufacturers. The three medical associations and the equipment manufacturers controlled the membership of this Committee until after World War II.

In 1946 the name of the organization was changed to National Committee on Radiation Protection and Measurements and its membership and programs were greatly expanded. In 1964 a congressional charter was issued to this organization and its name again changed to the National Council on Radiation Protection and Measurements. The charter allowed for it to become a self-perpetuating advisory board with considerably enhanced status. At the present time, there are 65 members of the NCRP, plus approximately 150 additional participants serving on its 36 scientific committees. Dr. Taylor[281] has indicated that the current NCRP membership consists of 15 health physicists, 11 radiological physicists, 8 physicists specializing in radiation measurement rather than radiation protection, 10 physicians, 17 radiobiologists, and 11 persons classified in other categories (this total exceeds 65, since some members are classified in more than one group). The majority of these NCRP members come from universities.

The NCRP, then, is a private organization of scientific experts specializing in radiation science that formulates recommendations for radiation control. The Committee is purely advisory, and its recommendations carry no legal status. However, they are usually adopted by state and federal agencies and codified. The programs and activities of the NCRP are designed to:[282]

1. Collect, analyze, develop and disseminate in the public interest, information and recommendations about:

a. protection against radiation

b. radiation measurements, quantities and units, particularly those concerned with radiation protection.

2. Provide a means by which organizations concerned with the scientific and related aspects of radiation protection and of radiation quantities, units, and measurements may cooperate for effective utilization in their combined resources and to stimulate the work of such organizations.

3. Develop basic concepts about radiation quantities, units, and measurements, about the application of these concepts and about the radiation protection.

4. Cooperate with the International Commission on Radiological Protection, the Federal Radiation Council, the International Commission on Radiation Units and Measurements and other national and international organizations – governmental and private concerned with radiation quantities, units and measurements and with radiation protection.

Until recently, the NCRP issued its recommendations in the form of handbooks that were issued by the National Bureau of Standards (NBS). The first recommendations appeared as NBS Handbook No. 15 in 1931.[70] Now the reports are published, issued, and sold by the NCRP. Table 10 contains a listing of the 39 reports that have been published to the time of this writing and their year of publication. The reports that are of particular concern to this review are numbers 33, 34, 35, and 39. NCRP Report No. 39[282] contains the most recent recommendations regarding acceptable levels of exposure to ionizing radiation. Reports 33-35 are concerned with the application of radiation producing machines in diagnostic Roentgenology.

The NCRP recommendations that are most widely known and understood are those dealing with numerical levels of MPD. It is obvious that the task of establishing rather arbitrary acceptable levels of exposure to radiation is most difficult. However, perhaps even more difficult is the interpretation of the meaning of MPD. The earliest recommendations[70,72] employed the term, "tolerance dose", which implied that if one's activities were conducted while not exceeding this dose level, there was no risk of injury. This interpretation was not spelled out, but was inferred from the use of the term, "tolerance." For example, the 1937 recommendations of the International Commission on Radiological Protection[69] (ICRP) known at that time as the International X-Ray and Radium Protection Commission, stated:

"...the evidence at present available appears to suggest that under satisfactory working conditions, a person in normal health can tolerate exposure to X-rays or radium gamma rays to an extent of about 0.2 international Roentgens (R) per day, or 1.0 R per week. On the basis of continuous irradiation, during a working day of seven hours, this figure corresponds to a tolerance dosage rate of 10^{-5} R per second. The protective values given in these recommendations are generally in harmony with this figure under average conditions."

The NCRP agreed with this tolerance dose until its recommendations of 1938,[283] which stated:

"...in the absence of precise lead protective data at the higher voltages each installation should, upon completion, be carefully tested for stray radiation which should not exceed the tolerance dose (0.1 R per day) at any possible position of the operating personnel.

Any deficiencies in protection should be rectified before routine operation commences. Tests for scattered radiation shall be made with an ionization instrument. The tolerance dose of 0.1 R per day stated above is provisional and it is advisable to apply generous safety factors. The tolerance dose may be measured with an

TABLE 10

Reports of the National Council on Radiation Protection and Measurements and Their Year of Publication

NCRP Report No.	Year	
1	1931	X-ray Protection
2	1934	Radium Protection
3	1936	X-ray Protection
4	1938	Radium Protection

Reports of the National Council on Radiation Protection and Measurements and Their Year of Publication

NCRP Report No.	Year	
5	1941	Safe Handling of Radioactive Luminous Compounds
6	1949	Medical X-ray Protection up to Two Million Volts
7	1949	Safe Handling of Radioactive Isotopes
8	1951	Control and Removal of Radioactive Contamination in Laboratories
9	1951	Recommendations for Waste Disposal of Phosphorus-32 and Iodine-131 for Medical Users
10	1952	Radiological Monitoring Methods and Instruments
11	1953	Maximum Permissible Amounts of Radioisotopes in the Human Body and Maximum Permissible Concentrations in Air and Water
12	1953	Recommendations for the Disposal of Carbon-14 Wastes
13	1954	Protection Against Radiations From Radium, Cobalt-60 and Cesium-137
14	1954	Protection Against Betatron-Synchroton Radiation up to 100 Million Electron Volts
15	1953	Safe Handling of Cadavers Containing Radioactive Isotopes
16	1954	Radioactive Waste Disposal in the Ocean
17	1954	Permissible Dose From External Sources of Ionizing Radiation Including Maximum Permissible Exposure to Man
18	1955	X-ray Protection
19	1955	Regulation of Radiation Exposure by Legislative Means
20	1957	Protection Against Neutron Radiation up to 30 Million Volts
21	1958	Safe Handling of Bodies Containing Radioactive Isotopes
22	1959	Maximum Permissible Body Burdens and Maximum Permissible Concentrations of Radionuclides in Air and in Water for Occupational Exposure
23	1960	Measurement of Neutron Flux and Spectra for Physical and Biological Applications
24	1960	Protection Against Radiations from Sealed Gamma Sources
25	1961	Measurement of Absorbed Dose of Neutrons and of Mixtures of Neutrons and Gamma Rays
26	1961	Medical X-ray Protection Up to Three Million Volts
27	1961	Stopping Powers for Use with Cavity Chambers
28	1961	A Manual of Radioactivity Procedures
29	1962	Exposure to Radiation in an Emergency
30	1964	Safe Handling of Radioactive Materials
31	1964	Shielding for High-Energy Electron Accelerator Installations
32	1966	Radiation Protection in Educational Institutions
33	1968	Medical X-ray and Gamma-ray Protection for Energies up to 10 Mev - Equipment Design and Use
34	1970	Medical X-ray and Gamma-ray Protection for Energies Up to 10 Mev - Structural Shielding Design and Evaluation
35	1970	Dental X-ray Protection
36	1970	Radiation Protection in Veterinary Medicine
37	1970	Precautions in the Management of Patients Who Have Received Therapeutic Amounts of Radionuclides
38	1971	Protection Against Neutron Radiation
39	1971	Basic Radiation Protection Criteria

Most of these reports are available for a nominal charge from NCRP Publications, P. O. Box 4867, Washington, D. C. 20008.

ionization chamber or photographic film placed at all possible positions of operators."

Interestingly enough, as Stone[27] has pointed out, this reduction by a factor of 2, from 0.2 R/day to 0.1 R/day, was done without any written explanation.

Taylor,[75] in commenting on the term, tolerance dose, in 1941, said, "The so-called tolerance dose is the total X-ray energy that a person may receive continuously without suffering any damage to the blood or reproductive organs." Henshaw's[136] interpretation of the tolerance dose was stated as follows:

"At the present stage in the history of protection, it is clear that the idea of tolerance dose automatically implies threshold type reaction, that is, that there must exist so far as the injurious effects are concerned, doses below which no effect is produced."

With evidence mounting that only acute radiation responses were threshold type and that long-term somatic and genetic consequences were non-threshold type, the concept of tolerance dose was in for revision.

The revision consisted of replacing the term, "tolerance dose," by the term, "maximum permissible dose." Many feel that this term is equally objectionable, but at least the concept has been more rigorously spelled out. Several authors[27,39,136,284] have reviewed the evolution of this term and its numerical value. All agree on its rather arbitrary nature, but then, in fact, so does the NCRP[285] when it states:

"The concept of a tolerance dose involves the assumption that if the dose is lower than a certain value–the threshold value–no injury results. Since it seems well established that there is no threshold dose for the production of gene mutations by radiation, it follows that strictly speaking, there is no such thing as a tolerance dose when all possible effects of radiation on the individual and future generations are included. In connection with the protection problem, the expression has been used in a more liberal sense, namely, to represent a dose that may be expected to produce only "tolerable" deleterious effects if any are produced at all. Since it is desirable to avoid this ambiguity, the expression "permissible dose" is much to be preferred.

"It is now necessary to give this expression a more precise meaning. In the first place, it is well to state explicitly that the concept of a permissible dose envisages the possibility of radiation injury manifestable during the lifetime of exposed individuals or in subsequent generations. However, the probability of the occurrence of such injuries must be so low that the risk would be readily acceptable to the average individual. *Permissible dose may then be defined as the dose of ionizing radiation that in the light of present knowledge, is not expected to cause appreciable bodily injury to a person at any time during his lifetime.* (NCRP italics)."

6.3 Maximum Permissible Dose Equivalent

Values of MPD are expressed in rem and the current recommendations, as developed by the NCRP,[282] are shown in Table 11. This table of dose limiting values differs considerably from the earliest recommendations. In addition to a value for the whole body of occupationally exposed persons, values are also recommended for certain organs, anatomical regions, and human health conditions. In addition, MPD's are specified for the general public under certain conditions, and for individuals under emergency conditions.

Although these dose limiting recommendations are incorporated into most regulatory statutes, their application must be tempered by the following qualifying statements of the NCRP:[282]

"At the present time, no inflexible numerical criteria governing radiation exposure can be given. (NCRP italics). Numerical limits to ranges of permissible dose equivalents have to be offered as guidance with the understanding that interpretation may vary, sometimes widely, from situation to situation. Modifications of the maximum permissible values recommended here may be justified because of social, technical and economic factors, provided they are designed in a competent fashion, are properly made known to those affected by the modifications, and conform with the general principles set forth in this report. As a general guide, all radiation exposure should be held to the lowest practical level, both as to dose equivalent, and dose equivalent accumulation rate."

It should be emphasized that these recommended MPD's apply only to occupational exposure. They do not include and should not be modified by considerations for exposure to natural background radiation or man-made devices outside the occupational environment, such as fall-out from nuclear weapons, environmental contributions from nuclear power plants, color TV sets, and shoefitting fluoroscopes (where they might still exist). Nor do these recommendations include exposures received as a patient during medical and dental procedures.

The basic protection criterion is that the accumulated dose equivalent to the whole body, which includes not only the whole body, but also partial body exposure to the gonads, the lens of the eye, or to portions of the red bone marrow shall not exceed 5 rems multiplied by the number of years of age beyond 18. Mathematically, this is expressed:

$$\text{MPD (total accumulated)} = (N\text{-}18) \times 5 \text{ rems}$$

TABLE 11

Maximum Permissible Dose Equivalent for Various Exposure Conditions

Condition	Dose Equivalent
Occupational Exposure	
Whole body	
Annual limit	5 rems
Long term accumulation to age N years	(N - 18) x 5 rems
Skin - Annual limit	15 rems
Hands - Annual Limit	75 rems
Quarterly Limit	25 rems
Forearms - Annual Limit	30 rems
Quarterly Limit	10 rems
Other organs, tissues and organ systems	15 rems
Fertile women in gestation period (with respect to fetus)	0.5 rem
The Public, or Occasionally Exposed Individuals	
Individual or Occasional-Annual limit	0.5 rem
Students-Annual Limit	0.1 rem
Population Limits	
Genetic-Annual Limit	0.17 rem average
Somatic-Annual Limit	0.17 rem average
Emergency Limits-Life Saving	
Individual (older than 45 years if possible)	100 rems
Hands and Forearms	200 rems, (300 rems, total)
Emergency Limits-Less Urgent	
Individual	25 rems
Hands and Forearms	100 rems total
Family of Radioactive Patients	
Individual (under age 45) Annual Limit	0.5 rem
Individual (over age 45) Annual Limit	5 rems

From NCRP Report No. 39.[282]

where N is the age in years and is therefore greater than 18. Simply put, the MPD for radiation workers is 5 rem per year which reduces to the administratively convenient value of 100 mrem per week. These are the values around which an efficient radiation control program should be developed and in actual practice, most diagnostic Roentgenological control programs rarely exceed one fifth of these values. However, should it be necessary for some workers to exceed this MPD level, then they should not be allowed to receive more than 15 rem in a given year, and only if their total accumulated MPD is not exceeded.

The MPD for unlimited areas of the skin is 15 rem per year. This recommendation is appropriate for exposure to radiations of low penetrability, such as electrons and low energy photons. The MPD for the hands is 75 rem per year of which no more than 25 rem may be received in any given calendar quarter. Similarly, the MPD for the forearms is 30 rems per year, of which no more than 10 rems may be received in a calendar quarter. These last two recommended levels carry particular significance to diagnostic radiologists. As increasingly more special procedures are performed, some under absolute aseptic conditions which are incompatible with the use of protective lead gloves, the hazard due to upper extremity exposure of the radiologist may increase. Several authors[286,287] have pointed out that during general radiological practice, the restrictive exposure to the radiologist may be to his hands and forearms, rather than to any other portion of the body, the higher MPD's notwithstanding. For any other tissue or organs, the MPD is 15 rem per year.

Because of the accumulating evidence on the increased radiosensitivity of the fetus to radiation, the MPD for an expectant mother is 0.5 rem for the entire gestation period. In applying this recommendation, it should be kept in mind that the radiosensitivity of the fetus is apparently greatest during the early stages of pregnancy and becomes less as gestation progresses. As Williams[288] and Langmead[289] recently pointed out, this does not mean that a female radiologist or technician should be removed from the clinic following the determination of pregnancy. On the basis of exposures normally received by these persons (see Table 20), work restrictions may not be necessary at all. On the other hand, if the clinical support staff is of sufficient size to allow the assignment of pregnant workers to low exposure procedures, this should be done. It would be prudent if pregnant workers were not allowed to participate in routine fluoroscopy and special procedures. If differences exist in the construction characteristics of Roentgenographic control booths, then pregnant workers should be assigned to those stations providing the most protection. Although this may seem stretching a point, evidence is available[290] to indicate that even minor design differences in control booths can contribute to reduced occupational exposure. Under some circumstances, additional personnel monitoring devices for pregnant personnel might be in order.

MPD's for the general public are specified for both individuals in the population and the population at large. It should be emphasized that these are MPD's for non-occupationally exposed persons where the exposure arises from sources not including natural background radiation and medical and dental exposures received as a patient. The MPD's for any individual in the population is 0.5 rem per year. This limit is based primarily on possible somatic effects. As the NCRP has stated, however:[282]

> "In an appraisal of radiation protection criteria from first principles, it would seem better to attempt to set limits for members of the public so that the risk of unacceptable injury to any individual is small compared with the total risk arising from any other environmental practice. Any such attempt fails, however, because the risks lack a common scale and few are known as well as those from radiation."

The MPD recommended for the population at large is 170 mrem per year. This level is based upon considerations of both somatic and genetic effects in the population and does not include natural background radiation and radiation received as a patient. Although this recommendation has been criticized recently,[181] the NCRP has stated[282] that this

> ". . .formulation is somewhat more restrictive than the NAS-BEAR recommendation of 1956[291] in spite of the fact that radiation research since then indicates that the genetic damage is probably less than then calculated. The continuation of this conservative position is justified by the residual uncertainties in translating data from animals to man, and by the demonstrated practicability of the limit."

Maximum permissible dose equivalents have also been recommended for emergency situations. Under the most urgent circumstances, whole body dose equivalents up to 100 rem are acceptable while the upper extremities may receive an additional 200 rem. The NCRP states that these recommendations are offered only as guidance and qualifies these levels with the following statements:[282]

> "It is compatible with the risk concept to accept exposures leading to doses considerably in excess of those appropriate for lifetime use when recovery from an accident or major operational difficulty is necessary. Saving of life, measures to circumvent substantial exposures to population groups or even preservation of valuable installations may all be sufficient cause for accepting above normal exposures. Dose limits cannot be specified. They should be commensurate with the significance of the objective, and held to the lowest practicable level that the emergency permits."

Application of these emergency recommendations will rarely, if ever, be encountered in diagnostic radiology.

One recommendation that often is of concern in diagnostic radiology is the employment of persons under 18 years of age. Fulltime employment of persons under 18 years of age in a radiation occupation should be avoided whenever possible. When these persons are employed, the MPD applicable for this group is 0.5 rem per year. If persons under 18 years of age are in an educational program or training program that is not considered employment, then they should not receive a dose equivalent exceeding 0.1 rem per year. If the educational program involves employment, such as a work-study program characteristic of many x-ray technician schools, the MPD for these individuals would still be 0.5 rem per year and not 0.6 rem per year.

6.4 International Commission on Radiological Protection

The ICRP was organized by the International Congress of Radiology in 1928 as the International X-Ray and Radium Protection Commission and was the model organization from which the NCRP was patterned. Taylor was involved in its organization and has described its history.[292] Its name was changed in 1950, when it assumed its current organizational structure. The ICRP is an arm of the International Congress of Radiology and it is composed of a chairman and not more than 12 members who are selected by an International Executive Committee on the basis of their excellence in the fields of radiology, radiation protection, physics, biology, genetics, biochemistry, and biophysics. Members generally serve for a term spanning three meetings of the Congress. The program of the commission has been spelled out as follows:[293]

> The policy adopted by the ICRP in preparing its recommendations is to deal with the basic principles of radiation protection, and to leave to the various national protection committees the right and responsibility of introducing the detailed technical regulations, recommen-

dations, or codes of practice best suited to the needs of the individual countries.

ICRP recommendations have been published in booklet form by the Pergamon Press since 1960. Prior to that time, their recommendations were published in various scientific journals and some of these publications have been referred to previously. Table 12 contains a listing of the ICRP publications, many of which are available at nominal cost.

Attention is directed to the latest publication, ICRP No. 16, entitled Protection of the Patient in X-Ray Diagnosis. This document should be reviewed by every radiologist and be part of his professional library.

6.5 International Commission on Radiation Units and Measurements

The International Commission on Radiation Units and Measurements (ICRU) was formed to develop internationally acceptable recommendations regarding:[278]

1. Quantities and units of radiation and radioactivity.
2. Procedures suitable for the measurement and application of these quantities in clinical radiology and radiobiology.
3. Physical data needed in the application of these procedures, the use of which tends to assure uniformity in reporting.

There are 13 members on the commission and some 140 persons from 16 different countries serving on task groups and sub-groups. Taylor has also reviewed the history of this organization.[294]

The ICRU, like the ICRP, is an arm of the International Congress of Radiology. It was organized at the first meeting of the Congress in 1925 and has been active ever since. In addition to its relationship with the Congress, the ICRU has

TABLE 12

Recommendations of the International Commission on Radiological Protection

	Recommendation	Year	Source
1.	X-ray and Radium Protection	1928	*Brit. J. Radiol.*, 1, 359, 1928.
2	Alterations to the 1928 recommendations	1931	*Brit. J. Radiol., 4, 485, 1931.*
3.	Revised recommendations	1931	*Brit. J. Radiol.*, 5, 82, 1932.
4	Revised recommendations	1934	*Brit. J. Radiol.*, 7, 695, 1934.
5.	Revised recommendations	1937	*Brit. Inst. Radiol.*, 1938.
6.	Revised recommendations	1950	*Brit. J. Radiol.*, 24, 46, 1951.
7.	Revised recommendations	1954	*Brit. J. Radiol.*, Suppl. 6, 1955.
8.	Amendments to recommendations	1956	*Radiology*, 70, 261, 1958.
9.	Recommendations of the ICRP. Publ. No. 1.	1959	Pergamon Press
10.	Permissible Dose for Internal Radiation Publ. No. 2	1959	Pergamon Press
11.	Protection against x-rays up to energies of 3MeV and Beta and Gamma Rays from sealed sources. Publ. No. 3	1960	Pergamon Press.
12.	Recommendations of the ICRP Publ. No. 6	1962	Pergamon Press
13.	Protection against electromagnetic radiation above 3MeV and Electrons, Neutrons and Protons. Publ. No. 4		Pergamon Press
14.	Report of the RBE Committee	1963	*Health Phys.*, 9, 357, 1963.

Recommendation	Year	Source
15. Handling and disposal of radioactive materials in Hospitals and Medical Research Establishments. Publ. No. 5	1965	Pergamon Press
16. A review of the physiology of the gastrointestinal tract in relation to radiation doses from radioactive material.	1966	*Health Phys.*, 12, 131, 1966.
17. Deposition and retention models for internal dosimetry of the human respiratory tract.	1966	*Health Phys.*, 12, 173, 1966.
18. Radiobiological aspects of supersonic transport.	1966	*Health Phys.*, 12, 209, 1966.
19. Calculation of radiation dose from protons and neutrons to 400 MeV.	1966	*Health Phys.*, 12, 227, 1966.
20. Principles of Environmental monitoring related to the Handling of Radioactive Materials.	1966	Pergamon Press
21. The Evaluation of Risks from Radiation.	1966	Pergamon Press
22. Recommendations of the ICRM (adopted in 1965).	1966	Pregamon Press
23. Evaluation of Radiation Doses to Body Tissues from Internal Contamination due to Occupational Exposure.	1967	Pergamon Press
24. Radiosensitivity of the Tissues in Bone	1967	Pergamon Press
25. General Principles of Monitoring for Radiation Protection of Workers	1968	Pergamon Press
26. Radiation Protection in Schools for Pupils up to the Age of 18 years	1968	Pergamon Press
27. Radiosensitivity and Spatial Distribution of Dose	1968	Pergamon Press
28. Protection Against Ionizing Radiation from External Sources	1969	Pergamon Press
29. Protection of the Patient in X-ray Diagnosis	1969	Pergamon Press

Recent publications are available from Pergamon Press, Maxwell House, Fairview Park, Elmsfor, New York 10523

official organizational ties to the World Health Organization (WHO), the International Atomic Energy Agency (IAEA), the United Nations Scientific Committee on the Effects of Atomic Radiation (UNSCEAR), and it has close, but unofficial, ties to many other organizations.

Since 1967, the ICRU has published and distributed its own recommendations in the form of soft-bound handbooks. Prior to that, the ICRU recommendations were published either as National Bureau of Standards Handbooks, or as scientific journal articles. Table 13 is a listing of these ICRU publications. It will be noted that most of the publications deal with subjects of interest primarily to radiation physicists. However, report No. 15, entitled Cameras for Image Intensi-

TABLE 13

Reports of the International Commission on Radiological Unites and Measurements

Report No.	Year	Title	Availability
1	1927	Discussion on International Units and Standards for X-ray Work	*Brit. J. Radiol.*, 23, 64, 1927
2	1928	International X-ray Unit of Intensity	*Brit. J. Radiol.*, 1, 363, 1928
3	1934	Report of Committee on Standardization of X-ray Measurements	*Radiology*, 22, 289, 1934
4	1934	Recommendations of the International Committee for Radiological Units	*Radiology*, 23, 580, 1934
5	1937	Recommendations of the International Committee for Radiological Units	*Radiology*, 29, 634, 1937
6	1951	Report of the ICRU	National Bureau of Standards Handbook 47
7	1954	Recommendations of the International Committee for Radiological Units	*Radiology*, 62, 106, 1954
8	1956	Report of the ICRU	National Bureau of Standards Handbook 62
9	1959	Report of the ICRU	National Bureau of Standards Handbook 78
10a	1962	Radiation Quantities and Units	National Bureau of Standards Handbook 84
10b	1964	Physical Aspects of Irradiation	National Bureau of Standards Handbook 85
10c	1963	Radioactivity	National Bureau of Standards Handbook 86
10d	1963	Clinical Dosimetry	National Bureau of Standards Handbook 87
10e	1963	Radiobiological Dosimetry	National Bureau of Standards Handbook 88
10f	1963	Methods of Evaluation Radiological Equipment and Materials	National Bureau of Standards Handbook 89
11	1968	Radiation Quantities and Units	ICRU
12	1968	Certification of Standardized Radioactive Sources	ICRU
13	1969	Neutron Fluence, Neutron Spectra and Kerma	ICRU
14	1969	Radiation Dosimetry: X-rays and Gamma Rays with Maximum Photon Energies Between 0.6 and 50 MeV	ICRU
15	1969	Cameras for Image Intensifier Fluorography	ICRU
16	1970	Linear Energy Transfer	ICRU
17	1970	Radiation Dosimetry: X-rays Generated at Potentials of 5 to 150 kV	ICRU

Recent publications are available from: ICRU Publications, P. O. Box 4869, Washington, D. C. 10008

fier Fluorography, shows that ICRU reports may also be of direct interest to the diagnostic radiologist.

6.6 Other Advisory Groups

Other recommendations are occasionally developed by scientific societies or governmental groups and many are of interest to the diagnostic radiologist. Table 14 identifies some of these recommendations by title and provides information on availability. The list is intended to be characteristic of information that is available and not to be exhaustive. In addition, the ICRP and the ICRU occasionally publish joint reports. Two of the more notable and pertinent reports[295,296] were entitled Exposure of Man to Ionizing Radiation Arising from Medical Procedures.

TABLE 14

Radiation Protection Recommendations By Various Governmental and Scientific Groups

Proposed by	Title	Date	Availability
American College of Radiology	A Practical Manual on the Medical and Dental Use of X-rays with Control of Radiation Hazards	1958	American College of Radiology 20 North Wacker Dr. Chicago, Illinois 60606, free
Industrial Medical Association	Recommended Safe Practices in the Use of Diagnostic X-ray Equipment	1963	Journal of Occupational Medicine 5-318-320, June 1963
British Ministry of Health	Code of Practice for the Protection of Persons Against Ionizing Radiation Arising from Medical and Dental Use	1964	Her Majesty's Stationery Office 423 Oxford Street London W. 1, $1.50
Indiana State Board of Health and Indiana University Medical Center	Low Dosage Medical Roentgenography	1964	Indiana State Board of Health, Indianapolis, Indiana, free
U. S. Public Health Service	Medical Uses of Radium and Radium Substitutes	1965	U. S. Public Health Service Public Inquiries Branch Washington, D. C. 20201 PHS Publication #999-RH-16, free
British Ministry of Health	Radiological Hazards to Patients	1966	Her Majesty's Stationery Office 423 Oxford Street London, W. 1, $0.50
U. S. Public Health Service	Radiation Protection for Podiatrist and Patient	1968	Superintendent of Documents, U. S. Government Printing Office, Washington, D. C. 20402, $0.25
U. S. Public Health Service	Radium and the Physician	1968	U. S. Public Health Service Public Inquires Branch Washington, D. C. 20201 PHS Publication # 1689, free
U. S. Public Health Service	Fast Film Exposure and Processing in Dental Radiography	1970	U. S. Public Health Service Public Inquiries Branch Washington, D. C. 20201, free
Advisory Committee for the Study of Professional Communications in Radiological Health	How to Protect Patient and Physician During X-ray Examinations. Installment 1: Effects of Radiation	1970	American Family Practice/GP 1 113-128, January 1970
Advisory Committee for the Study of Professional Communications in Radiological Health	How to Protect Patient and Physician During X-ray Examinations. Installment 2: Responsible Use of Diagnostic X-rays	1970	American Family Practice/GP 1 105-120, February 1970
Indiana State Board of Health and Indiana University Medical Center	Low Dosage Pediatric Roentgenography	1970	Indiana State Board of Health, Indianapolis, Indiana, free

7. RADIATION EXPOSURE AND DOSE IN DIAGNOSTIC RADIOLOGY

The significance of our current radiation protection practices may not be clear without an understanding of radiation exposures and radiation doses experienced in present day diagnostic radiology. Therefore, the previously discussed dose limiting recommendations and the section following, section 8, which deals with devices and procedures designed to limit exposure will be supplemented and complemented with the discussion following. There are three areas in today's practice of radiology that require a knowledge of exposure and dose. They are related to the output intensity of the x-ray machine, the dose equivalent to the radiological personnel, and the radiation dose received by the patient. Only with this type of information can a critical assessment of radiation control in one's own radiological facility be attempted. Fortunately, the literature is replete with information bearing on each of these areas.

7.1.1 Diagnostic X-Ray Machine Output

In the past, primary concern for x-ray machine output has been concentrated on the fluoroscope. This occurred because the fluoroscope was known to be the source of most radiation exposure received by radiological personnel and because it had been implicated in numerous acute radiation injuries. Concern for fluoroscopic output still exists, but increasing demand and utilization of radiographic procedures have resulted in increased attention being given to radiation control considerations of these machines. In recent years the protective equipment and apparatus employed by radiologists during fluoroscopy have served to significantly reduce exposures to these operating personnel.

Both radiographic and fluoroscopic machine output has been steadily decreasing over the years. This has resulted from the availability of better equipment and the recognition that adequate useful diagnostic information was possible with reduced radiation intensities. Additionally, concern for patient and operator has contributed to machine and accessory design resulting in reduced output intensities.

Until approximately 1940, the availability of adequate measuring devices for machine output was severely limited. Lauriston Taylor, for example, has related an incident that occured in 1929 that required him to design and build a portable radiation survey meter because none was available.[297] It happened that Dr. Taylor apparently received an exposure of 200 R from a newly installed 200 kVp water cooled x-ray machine. Although this constitutes a very serious exposure, because of the information on biological effects available at the time, he did not consider the exposure alarming. As Dr. Taylor has related:

> "Not realizing that this much whole body exposure might have been dangerous, I was quite relieved at its low value in comparison with some of the levels of exposure which were then being commonly administered in radiotherapeutic applications. Since I was not aware that one should become nauseated at this exposure, I was not nauseated either."

7.1.2. Fluoroscopic Output

Braestrup followed a survey of stray radiation in radiotherapy facilities[298] with a similar series of measurements in diagnostic departments[299] that were reported in 1942. He used portable ionization chambers of his own design to measure leakage, scattered, and primary radiation from several radiographic and fluoroscopic units. He reported that the exposure levels due to leakage radiation at a distance of one meter from the target of nine different x-ray tubes varied from a low of 3 mR/hr to a high of 515 mR/hr. Of the nine tubes tested four of them exceeded our present day recommendations of 100 mR/hr at one meter from the target.[300] Measurements of the output from 13 different vertical fluoroscopes showed exposure rates from 6.8 to 127.6 R/min at the tabletop. Measurements made on 11 horizontal and tilt type fluoroscopes showed that the tabletop output varied from 9.6 to 55.2 R/min. Of these 24 different fluoroscopes, only 3 would have met later NCRP recommendations that tabletop output should not exceed 10 R/min.[301] By today's more flexible standards of 3.2 R/mA min,[300] 6 of the fluoroscopes would have been acceptable. The average output for all of these units was 28.2 R/min and all measurements were made at a tube current of 5 mA and a tube potential of approximately 80 kVp. The primary beam intensity of 4 diagnostic x-ray units operated at approximately 80 kVp with 0.5 mmAl added filtration was 4.82 R/100 mAs.

Concern over the increasing availability and use of fluoroscopes by radiologists and nonradiologists alike prompted Bell[302] to examine the radiation output of his fluoroscope before it was replaced with a newer unit. His original unit was approximately 12 years old in 1940 when the measurements were made. When operated at 88 kVp, 3 mA and a target to tabletop distance of 25 centimeters, the tabletop intensity was found to be 30 R/min. During actual fluoroscopic examinations of the chest, the output varied from 33 to 134 R total exposure. Total exposures ranging from 148 to 396 R were measured during actual fluoroscopic examinations of the G. I. tract. Bell's replacement fluoroscope was designed for a 38 centimeter target to tabletop distance and when it was operated at 3 mA and a voltage range of 70 to 84 kVp, it produced an average output of 15 R/min at the tabletop.

A number of other reports[303,304] in the 1940's and 1950's pointed out the dangers of fluoroscopy, especially in untrained or inexperienced hands. Chamberlain[305] found that the tabletop output of 2 of his fluoroscopes varied from 8.8 R/min to 68.2 R/min under various normal operating conditions. Of the 18 conditions reported, 13 resulted in outputs exceeding 20 R/min at the tabletop. A radiation control survey of 117 fluoroscopes reported in 1952 contained rather alarming results.[306] The measurements were made with ionization chambers while the physician-operator set the operating conditions. The output of these fluoroscopes ranged from 2 R/min to 118 R/min at the tabletop with 21% of them measuring more than 30 R/min. In 1955 Kirsh[307] reported the results of measurements of tabletop exposure rate for 3 unfiltered fluoroscopes at the Veterans Administration Hospital, Hines, Illinois. The 3 output intensities were 15.7, 24.0, and 5.9 R/min. The addition of 2.0 mmAl filtration reduced the output to approximately 30% of its original value.

More recently there have been some large-scale, well designed, authoritative surveys of medical radiological equipment. Most extensive of these were the nationwide exposure and dose surveys conducted by the Bureau of Radiological Health, U. S. Public Health Service for the year 1964.[308] Although many of the data collected were obtained by individual inspection and evaluation of a given radiological facility, many were also gathered through the use of the Medical Surpack.[309-312]*

The Medical Surpack is a mailable film package consisting of an x-ray film, a dosimeter film, and accessory devices. Its cost is approximately $20. Each Surpack contains three packets labeled "chest," "extremity," and "fluoroscope," along with instructions for their use. The Surpack is designed to measure the following parameters:

1. x-ray beam size
2. x-ray beam symmetry
3. half value layer
4. total filtration
5. x-ray intensity
6. fluoroscopic tube target to tabletop distance
7. beam collimation
8. fluorescent screen attenuation

In a preliminary report[313] of individual radiation control surveys of diagnostic equipment, the Public Health Service showed that although exposure levels from these machines were generally acceptable, other radiation control features were not. The survey covered 5263 x-ray facilities in 25 states and two territories during the period 1962 through 1967. Table 15 is extracted from this report and shows that only 73% of the fluoroscopes investigated conformed with the current NCRP recommendations for output intensity. Furthermore, 5% of the fluoroscopes examined had an intensity equal to or greater than 20 R/min at the tabletop. The data also indicate that the machines located in radiologists' offices were operated at lower x-ray intensities than those used by other specialists.

More recent studies,[314-316] conducted on a smaller scale, indicate that fluoroscopic output is becoming less of a radiation control problem. In a 1964 report covering 91.5% of the diagnostic x-ray equipment in Polk County, Florida (118 facilities), inadequate filtration, resulting in excessive x-ray intensity, was detected and corrected in 23 units. No excessive x-ray intensity was found in a comprehensive survey of the x-ray facilities of the Bureau of Prisons.[315] The findings resulted from radiation control surveys of 27 facilities containing a total of 17 fluoroscopes. A similar finding of no excessive x-ray intensity in fluoroscopes resulted from radiation control surveys in 11 Public Health Service hospitals and 20 outpatient clinics throughout the United States.[316] These surveys were performed between March 1968, and February 1969.

*(Any radiologist wishing to participate in the Medical Surpack program should communicate directly with the Bureau of Radiological Health, Environmental Protection Agency, Washington, D. C.)

TABLE 15

Fluoroscopic Intensity at the Table-top by Type of Medical Practice. Each value represents the percent of machines operating in the given output range. (Roentgens/min)

Roentgens/min	Radiologists	Internal Medicine	Pediatricians	General Practice	Osteopaths	Chiropracters	Total*
<1.0	1.1	0.9	-	1.3	-	-	1.5
1.0 - 9.9	89.5	59.4	58.5	60.0	32.4	45.0	71.8
10.0 - 14.9	5.5	15.4	17.1	7.7	14.7	5.0	7.4
15.0 - 19.9	-	6.5	7.3	4.5	5.9	10.0	3.2
>20.0	1.2	10.3	14.7	8.2	8.8	5.0	4.8
Unknown	2.8	7.5	2.4	18.3	38.2	35.0	11.4
Percent in Compliance with NCRP #26	90.6	60.3	58.5	61.3	32.4	45.0	73.3

*There are other reported specialties not reported in this table that contribute to the total values.
From Fess and Seabron.[313]

7.1.3. Radiographic Output

Concern for the x-ray output of radiographic tube-head has not been obvious until recently and, even now, the primary consideration for radiographic tubes is directed to filtration and collimation rather than to a precise determination of output intensity. Recommendations for maximum fluoroscope output are published because of the high intensity time required for each examination. No comparable recommendations are available for radiographic equipment.

Still, it is of interest at times to be able to estimate the output of the radiographic head for certain conditions of operation. The most straightforward way of doing this is by direct measurement. However, many radiologists do not have access to proper radiation measurement equipment and, therefore, the use of charts or nomograms may be necessary.

Figure 1 is a nomogram based on measurements by Osborn and published by the ICRP.[317] This figure allows one to estimate the average exposure to the skin in mR per mAs for fully rectified diagnostic x ray tubes operated between 40 and 150 kVp. Two conditions of filtration are covered, 2 mmAl and 4 mmAl, and the nomogram is good for focus to skin distances from 10 to 80 inches. To use the nomogram one needs know only the kVp of operation, the approximate filtration, and the focus to skin distance. The exposure in mR/mAs is determined at the intersection of the central column and a straight line between the appropriate operating points on the outside columns. It should be emphasized that because of considerable differences in x-ray tube design, this nomogram provides only approximate values and, therefore, differences from the true value may be as much as ± 30%.

McCullough and Cameron[318] have published two useful figures for estimating x-ray exposure from a diagnostic unit as a function of filtration and kVp. These relationships are shown in Figures 2 and 3 for examination distances of 24 and 40 inches, respectively. The graphs are based on computations and the authors indicate that their results are in good agreement with experimental values over clinically employed ranges of filtration and kVp.

Jackson[319] has also related a method for assessing radiographic output and patient dose to critical organs. Although his approach does result in considerable useful data, it is not very straightforward, and the reader is referred to the original article for details.

7.2.1. Occupational Exposure in Diagnostic Radiology

The photographic film badge has long been the standard methodology for estimating radiation exposure to diagnostic personnel. Although its use was first suggested by Pfahler in 1922,[44] the film badge did not become generally accepted by radiologists until the late 1940's. This acceptance followed the experience with film badge monitoring developed on the Manhattan Project.[27]

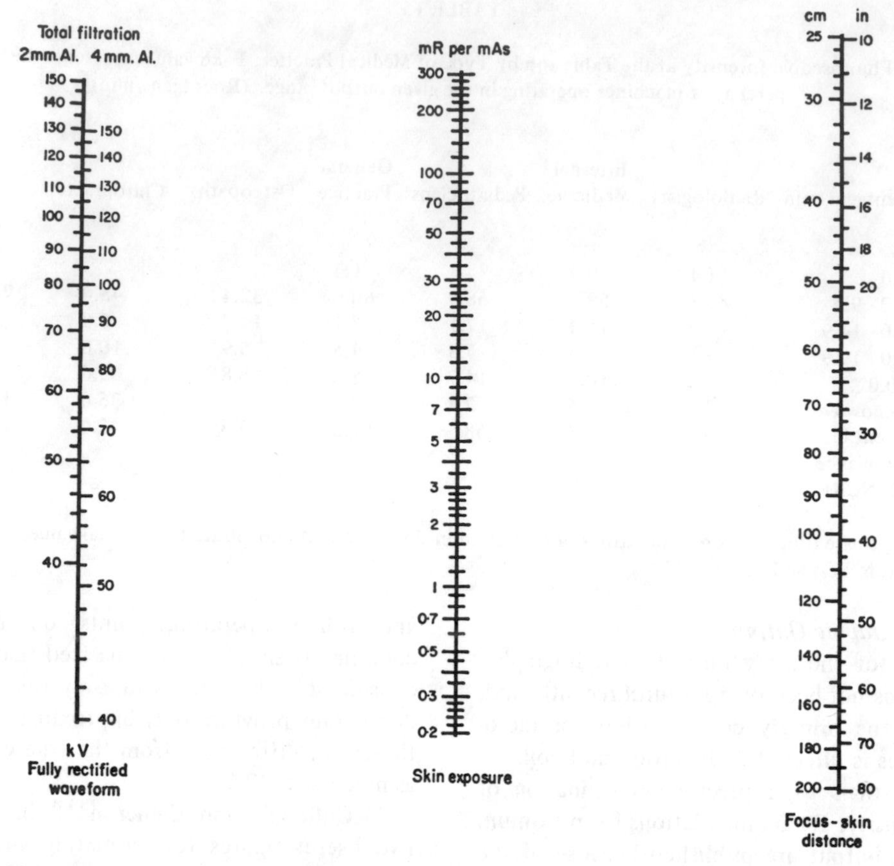

Figure 1. A nomogram for determining the approximate radiation intensity produced by a radiographic unit under variable operating conditions. A straight edge through appropriate points on the outside scales will intercept the middle scale at the approximate skin exposure value. From ICRP Publication 16.

7.2.2 Some Limitations of Personnel Monitoring

The technical aspects of film badge personnel monitoring have changed very little from the early review by Wilsey.[320] Film badges have proven to be relatively accurate and reliable as recent quality control surveys of commercial suppliers of film badges have shown.[321,322] At one time, pocket ionization chambers were used in addition to, or in place of, film badges for personnel monitoring,[323] but these devices were found to be unsuitable for routine use in diagnostic radiology for a number of reasons. Currently the application of thermoluminescence dosimetry (TLD) to personnel monitoring is receiving widespread attention.[324–327] Most companies specializing in personnel monitoring services now offer some TLD devices in addition to their normal line of film badges. Thermoluminescence analyzers and materials are becoming available in the price range that makes this type of personnel monitoring accessible and compatible with an in-house project for some larger x-ray departments.

Personnel monitoring in diagnostic radiology has enjoyed reasonable usefulness over the past 20 years. However, very severe questions are being raised currently about its purpose and benefits. In addition to providing estimates of exposure to radiation of employees, personnel monitoring has also been reported effective in allowing detection of serious equipment malfunction[328] and in correcting certain inadequacies in individual fluoroscopic technique.[329,330] Perhaps the greatest benefit derived from personnel monitoring is making the personnel aware that although they cannot sense radiation, it is still present and should be treated with respect. The numerical evaluation of an individual's personnel monitor carries little significance to the individual even though some would attach considerable importance to the dose equivalent reported. Because of the many pitfalls

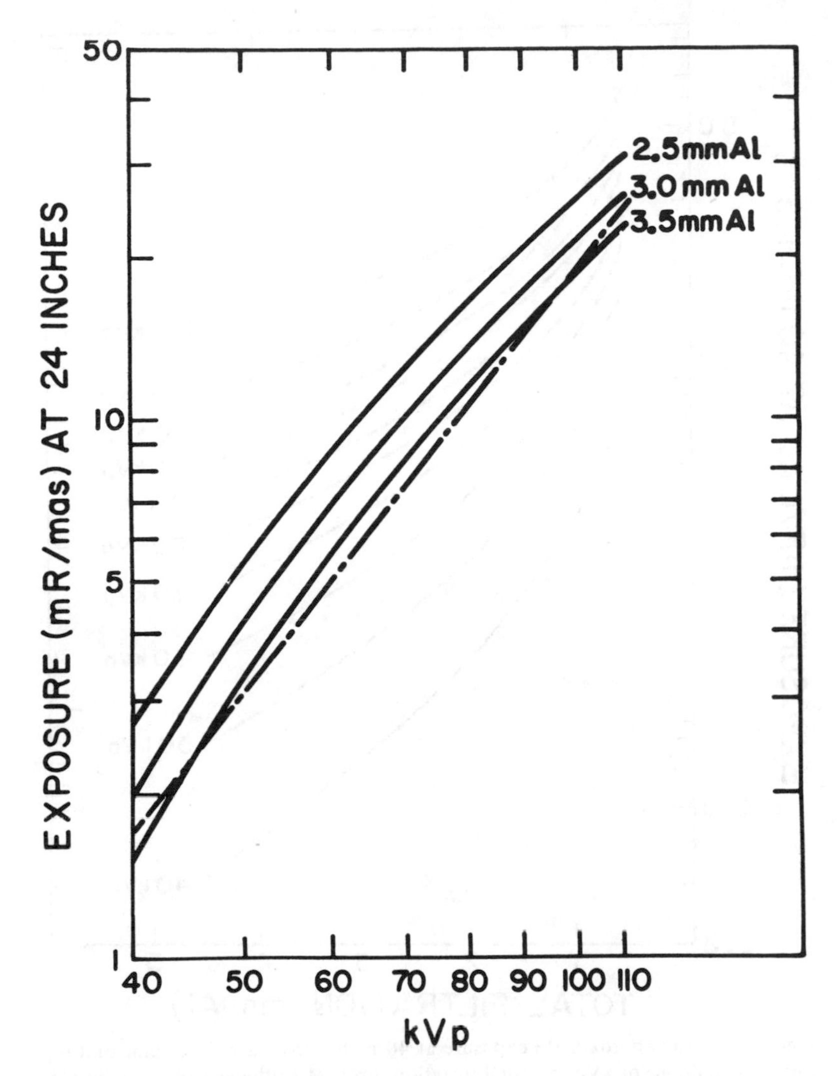

Figure 2. The approximate exposure at 24 inches from a radiographic unit for various conditions of kVp and total filtration. From McCullough and Cameron.[318]

associated with personnel monitoring, there is little, if any, relationship between the reported dose equivalent and any acute or latent biological response. In general terms these pitfalls relate to the anatomical location where the monitor is worn, the quality of radiation to which the individual is exposed, and the temporal and spatial distribution of that radiation exposure.

Personnel monitoring records are required to be accumulated for a 30-year period by British authorities and indefinitely by American authorities.[331,332] The reasons for these long-term records keeping responsibilities seem to be threefold. First, the records may be used in adjudication of workmen's compensation cases; secondly, it may be possible to use them in long-term epidemiology studies; and finally, application to demographical studies may be possible. These reasons have been rejected by at least one scientific society[333] and by several individual radiation scientists[334,335] as being insufficient justification for long-term radiation records keeping. This view was also repeatedly expressed at the Conference on Medical Radiation Information for Litigation held in 1968.[336] The Health Physics Society has identified the primary purposes to be served by a radiation monitoring program, none of which supports long-term radiation records keeping.[333]

1. To monitor the employee's radiation environment and to evaluate the ade-

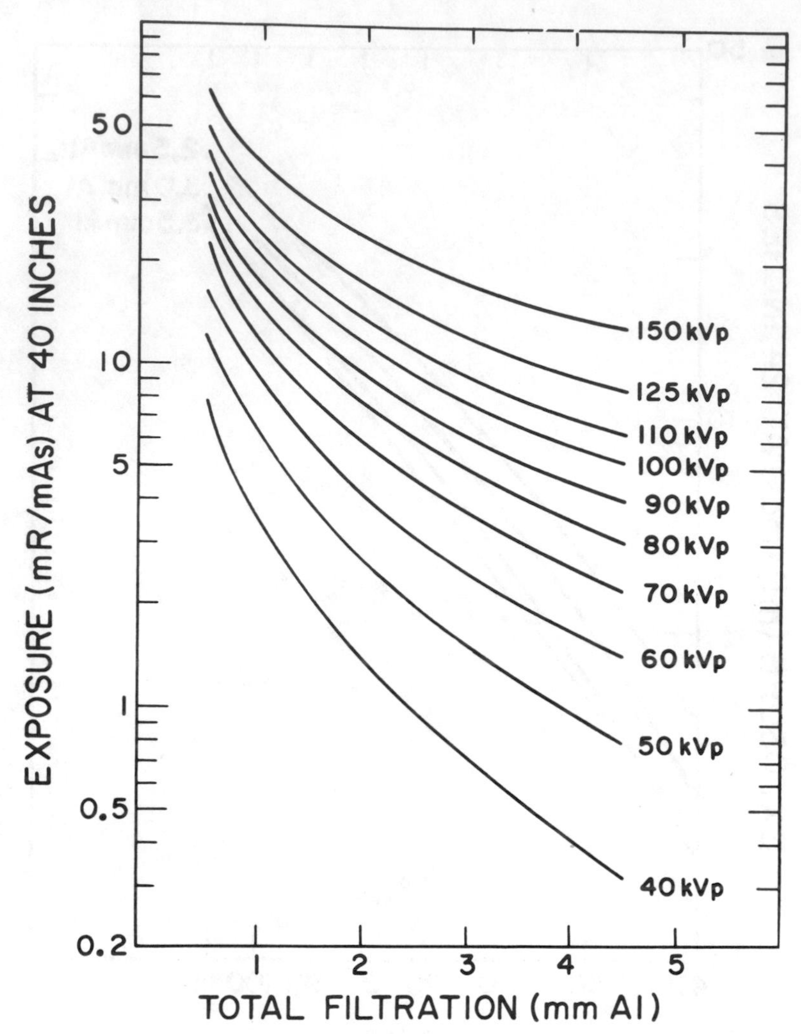

Figure 3. The approximate exposure at 40 inches from a radiographic unit for various conditions of kVp and total filtration. From McCullough and Cameron.[318]

quacy of the environmental radiation control program.

2. To promote safe radiation working habits on the part of individual employees.
3. To document radiation accidents.
4. To satisfy the employer's medico-legal requirements as are necessary to protect the employee and his employer.
5. To comply with pertinent federal, state, and local regulations and laws.

In contradiction to regulatory requirements, there has been considerable discussion in the recent literature with general agreement that there are severe limitations to personnel monitoring,[334] and even more questionable is the practice of pre-employment and periodic routine medical examination of radiation workers.[335,337,338] The practice of requiring pre-employment and routine medical examinations apparently stems from recommendations of advisory groups such as the ICRP. In Publication #9 the ICRP states: [339]

> "The assessment of health, both before and during employment, is directed towards determining whether the health of the worker is compatible with the tasks for which he is employed. The type and extent of the surveillance should be the same as in general industrial medical practice and should include both pre-employment and routine examinations, the frequency of the latter being determined mainly by the individual's general health and the conditions of work. . . ."

It is recommendations such as these that are currently being brought to task. Similar questions and concerns are raised regarding the maintenance of cumulative radiation dose records.[340,341] If the current trend of requiring more complete and long-term records continues, then, no doubt, the only tenable method of satisfying these requirements would be large-scale computerization of records.[342] An even more horrifying possibility is the establishment of a central registry for radiation doses received as a patient. This was proposed long ago[343] and Hodges[344] pointed out that federal legislation had been submitted in support of the central registry idea.

A perfect example of the limitations of personnel monitoring in diagnostic radiology is the difference in recorded dose equivalent depending upon where on the body the monitor is worn. During fluoroscopy when the normal radiation field is perturbed by protective apparel, the dose equivalent to various parts of the body will range over a wide value. This makes determination of the proper place to wear the monitor difficult. Recommendations in this regard are inconsistent and contradictory.

1. Ordinarily the best place to wear a monitoring device is on the chest or under the lead apron if one is worn.[345]

2. The proper place for a film badge when a lead apron is being worn is near the top of the apron and on the outside.[346,347]

3. It may be pointed out here that badges should be worn where the maximum radiation may be received. The radiologist who wears an apron should place the badge on the upper arm, on the forearm, or on the leg.[286]

4. The monitor should be worn on the trunk at the site most likely to be at risk, irrespective of whether or not it is sometimes covered by protective clothing. When all parts of the trunk are considered equally at risk, I recommend waist level as appropriate.[348]

5. . . .particular care should be taken in choosing the location of the monitoring device. . .accordingly, a qualified expert should be consulted.[331]

6. Obviously, if the film badge is worn on the chest and the radiation is always incident on the back, a meaningless record is obtained. The range of typical circumstances has to be reviewed in each work area, with the disposition of measuring devices made reasonable for the nature of the work.[282]

Recent papers on this subject[287,349] have indicated that in diagnostic radiology, the most appropriate place to wear the film badge, if only one is available, is on the front collar above the protective apron. These reports showed that this region of the body received a higher percent of the applicable MPD than in any other whole-body location. Critical evaluation of extremity monitoring was also pointed out as being necessary. Unquestionably, if only one film badge is worn, a notation of the position at which it is worn should be an integral part of the individual personnel monitoring record if that record is to be used for any long-term evaluations of individual or population response to radiation.

7.2.3. Exposure to Nonradiologists

Although this review is directed to radiologists, some attention must be directed to nonradiologists who employ radiation producing machines in their practice. Periodic reports[350-354] have cautioned nonradiologists about the hazards associated with using x-rays. The reports point out that nonradiologists are generally less aware of these hazards and less trained and motivated to accommodate them. Personnel monitoring by nonradiologist specialists, particularly those in private office practice, frequently does not exist. In the report by Leddy,[350] for example, from a total of 55 physicians who reported to the Mayo Clinic for treatment of Roentgen ray dermatitis, 44 of them were nonradiologists. The early practice by general physicians of reducing fractures under the fluoroscope was discontinued because of reports such as those indicating excessive extremity exposure.[351]

7.2.4. Spatial Exposure Distribution to Radiologists

Exposures to various parts of the radiologist's body during different fluoroscopic procedures have been reported. Generally, these reports have been concerned with protection of the fluoroscopist's upper extremities and with technique changes designed to reduce exposure to extremities as well as to the whole body. Measurements made during cystoscopic examination,[355,356] angiocardiography,[357,358] cardiac catheterization,[357] cholangiography,[359] examination of the stomach,[59] upper gastrointestinal tract examination,[329] and lower gastrointestinal tract examination[60,356] have all indicated the need to be aware of extremity exposures as well as the high level of scatter radiation to the trunk of the examining fluoroscopist. Consequently, not only are protective aprons necessary, but also protective gloves for both hands when the examination permits. When the examination does not permit both hands to be gloved, the smallest possible field sizes and most cautious technique are mandatory. In a survey of 15 fluoroscopy rooms Cowing and Spalding[360] noted that the smaller the room, the higher the level of stray radiation, and they suggested the possibility that the posterior exposure of radiologists might be greater than anterior exposure through the protective lead apron. Later investigations[287] showed this observation to be true. They recommended that the minimum room size should be at least 12 x 14 feet.

7.2.5. Dose Equivalents in Diagnostic Radiology

Through the years various estimates have been made of the ranges of personnel exposures experienced in radiology. Prior to the widespread use of film badges in the late 1940's, these measurements were conducted on a test basis usually to demonstrate that the recommended tolerance dose was not being exceeded. Pocket ionization chambers were most often employed and the earliest concern was given to radiologists working with radium.[50,361-364]

Once film badge personnel monitoring came into general use, articles reporting personnel exposures in diagnostic radiology were abundant, but as personnel exposures have decreased, so have these types of reports. It was pointed out in one of the early papers[119] that a totally unexpected side benefit from the adoption of routine personnel monitoring was the apparent creation of a psychological effect resulting in more careful personal technique. This finding was reported by Hunter et al.[119] from their experiences at the Massachusetts General Hospital as follows:

> "Although the monitoring has only been in effect since September, 1948 most personnel have shown progressively lower exposures. Should this happy trend continue, it may well be that the psychologic effect of having to wear a film badge will reduce exposure to such a degree that occasional blood examinations can eventually be substituted for those now done twice a month."

Their data showed that during the first year of routine personnel monitoring nearly all personnel had exposures less than 50 mrem per week.

Prior to the institution of routine personnel monitoring in diagnostic radiology, radiation measurements for radiation control purposes were generally confined to determinations of stray radiation levels.[365-367] Data generated during these types of investigations, in addition to analyses of the type of procedures performed, led Braestrup[368] to estimate that early radiologists received radiation exposures averaging approximately 100 R per year. This is a considerable exposure in light of today's experience. In this same discussion[368] Braestrup reports the results of film badge monitoring that had been conducted in a group of New York City Hospitals. From a total of 91 diagnostic radiologists the average annual dose-equivalent (DE) reported in 1957 was approximately 3000 mrem. In a 1942 study[299] evaluating film for personnel monitoring, Braestrup reported exposures to diagnostic personnel ranging from 0 to 200 mrem per month.

Other reports offered during this early period of film badge personnel monitoring indicated that considerable exposures were being received by diagnostic personnel. One of the earliest informative large-scale surveys employing photographic film as personnel monitors was reported by Clark and Jones[369] in 1943. An analysis of 2071 weekly film badges from four different hospitals showed that 27.3% of the badges received more than 250 mrem per week, and 9.2% of the badges received more than 500 mrem per week. Cowie and Scheele[370] reported that:

> "In 26 institutions the daily exposure received during Roentgenoscopy by at least one worker in each exceeded 0.1 R. The extreme was one institution in which the equipment provided little protection and the radiologist, who wore no apron, received 2 R per day on the body. Technicians wore aprons in only 7 of the 42 hospitals checked."

DeAmicis et al.[371] reported in 1952 that compilation of film badge monitoring data accumulated during one year from seven general hospitals. The film badges were changed either weekly or bi-weekly and of 1100 film badges worn by radiologists and 2740 worn by technicians, only 7 and 3 of the film badges, respectively, showed exposures exceeding the contemporary MPD of 300 mrem per week. Geist et al.[372] reported the personnel monitoring experience from the Cleveland Clinic Foundation in 1953. The survey covered a period of 3 months and involved 84 radiological personnel who wore film badges for weekly intervals. Some of the more interesting data reported in this study are summarized in Table 16. A report by Hunter and Robbins[120]

TABLE 16

Personnel Exposures in 1953 at the Cleveland Clinic Foundation for Various Radiological Activities

Activity	No. Personnel Monitored	Average* Dose Equivalent in m rem/wk
General Diagnostic Clinic (x-ray technicians)	12	<300 (except on 9 occasions)
General Diagnostic Clinic (physicians)	13	103
Hosptial Diagnostic Department (x-ray technicians)	9	97
Hospital Diagnostic Department (physicians)	7	60

*The MPD at that time was 300 m rem/wk.
From Geist et al.[372]

TABLE 17

Average Dose Equivalents (mrem/wk) to Personnel at Dunedin Hospital, New Zealand, During a Six-Year Period

Year	Number of Radiological Personnel	Average Personnel Exposure During Year - (mrem/wk)				
		<50	50-99	100-199	200-499	>500
1946	11	4	1	3	3	0
1947	13	7	3	3	0	0
1948	10	3	3	3	1	0
1949	12	4	4	3	1	0
1950	24	16	4	2	0	2
1951	26	18	5	2	1	0

From Jamieson.[375]

from the Massachusetts General Hospital showed that the average weekly exposure to individuals during the initial 20-month period beginning September 1948 ranged from background to 31 mrem per week.

Heustis and Van Farowe[373] reported that exposures in mental hospitals averaged less than 60 mR per week to technicians, but averaged approximately 1000 mR per week to various assistants who were required to hold the mental patients during examination. Osborn[374] reported the personnel monitoring experience from 35 persons employed at the University College Hospital in London. Individual exposures ranged as high as 700 mR per week. The average exposure for these persons was 56 mR per week and the data showed that the trained individuals generally received higher exposures than trainees.

Table 17 is abstracted from a report by Jamieson[375] and shows the personnel monitoring experience during a 6-year period at Dunedin Hospital in New Zealand. The MPD in effect during the time of this study was 300 mR per week. In all but one year covered by this survey at least one individual averaged greater than 200 mrem per week. Cowing[376] showed that between 1952 and 1957 the number of film badges with weekly radiation exposure greater than 100 mrem decreased from 9% in 1952 to 4% in 1957. These values are based on 10,000 film badges examined during 1952 and 12,500 examined during 1957. The data shown in Table 18 are extracted from an

TABLE 18

The Approximate Percentage of Personnel Exposures in Several Exposure Ranges for the 11-Year Period Beginning 1950 in 106 New England Hospitals

Year	Total No. Film Badges	<10 mR/wk	10-99 mR/wk	100-199 mR/wk	200-299 mR/wk	>300 mR/wk
1950	3747	74	20	4	1.6	.4
1951	4051	64	28	6	1.4	.6
1952	4427	61	30	7	1.3	.7
1953	5303	59	32	4	2.8	2.2
1954	5008	47	40	8	2.9	2.1
1955	6655	62	31	4	1.2	1.8
1956	6980	53	40	4	1.5	1.5
1957	7100	49	44	4	1.5	1.5
1958	7050	59	35	3	1.2	1.8
1959	6560	58	35	4	1.3	1.7
1960	7162	92	3	2	1.2	1.8

From Spalding and Cowing.[377]

article by Spalding and Cowing[377] who surveyed the personnel monitoring records in the x-ray departments of 106 New England hospitals between 1950 and 1960. These data indicate a generally constant level of exposure over the 11-year span of the study. Perhaps the increased personnel exposures expected from increasing radiological examinations during this period of time were offset by safer equipment development and greater attention to personal technique and procedures.

In 1970 Langmead and Steadman[378] reported dose equivalents for personnel in a large hospital in the northeast of England and for another hospital group in the south Midlands. Some of these data are shown in Table 19. Others[379] have presented some average annual DE values for several medical and nonmedical occupational groups monitored by the Radiological Protection Service. They showed, for example, that the average annual dose per worker was 0.35 rem for diagnostic radiology, 0.49 rem for radiotherapy, 0.54 rem for industrial radiography, 0.30 rem for crystallography, and 0.24 rem for veterinary radiography.

TABLE 19
Annual Personnel Dose Equivalent from Two Large Hospital Radiology Groups for a One-Year Period

	North-East England Hospital (1965-66)	South Midlands Hospital (1967)
Number of Monitored Personnel	205	205
<250 mrem	200	91
250-500 mrem	5	112
>500 mrem	0	2
average dose	50	<240
maximum dose	450	1730

From Langmead and Steadman.[378]

The data presented in Table 20 is from an unpublished analysis of personnel monitoring records at the four teaching hospitals in Houston, Texas, that are affiliated with the Baylor College of Medicine. Each is a large general hospital, with sizable departments of diagnostic radiology. The average dose equivalent values recorded were compiled from only those employees who were present during each given calendar year. If employment was restricted to part of the year, that individual was not included in that analysis. It is apparent that although there are slight differences in average exposures among the hospitals, the exposures are generally quite low when compared with reports of 20 years ago. As teaching hospitals, each of the radiology departments accommodates on a regular basis resident radiologists. Hospital A is the main teaching hospital and it generally has 12 or more residents on its staff. One might expect that lack of experience on the part of the residents would result in higher exposures. This, however, is not confirmed by the data in Table 20. This finding is also consistent with the data of Osborn[374] who showed that radiologists and technicians in training received less exposure than well trained individuals. The data reported in Table 20 does not include exposures received by

TABLE 20

Personnel Monitoring in Departments of Diagnostic Radiology in Four Baylor College of Medicine Affiliated General Hospitals Located in Houston, Texas

Number of Personnel Monitored/ Average Dose Equivalent (mrem)	1965	1966	1967	1968	1969	1970	Average annual D. E. per person (mrem)
Hospital A							
radiologists	------	7/309	16/454	10/361	8/427	9/333	389.1
technicians	------	25/421	31/324	41/306	49/563	48/214	365.9
Hospital B							
radiologists	9/446	8/339	8/314	7/275	8/543	8/718	442.5
technicians	44/316	52/187	50/155	55/185	72/208	24/414	140.2
Hospital C							
radiologists	2/238	4/267	6/393	6/1187	5/550	5/490	579.4
technicians	20/522	18/588	17/220	27/454	20/487	19/718	499.35
Hospital D							
radiologists	1/2010	3/457	5/593	6/1194	5/1750	3/2688	1318.4
technicians	8/487	5/510	6/397	9/937	15/942	10/971	775.5

TABLE 21

Workload Characteristics of the Four Hospital Radiology Departments Identified in Table 20. These data are for 1970.

	No. Beds	No. Radiology Patients	No. Films	Flouro and Special Procedures
Hospital A	476	106,141	354,102	10,342
Hospital B	1,040	72,858	98,394	13,702
Hospital C	669	60,112	180,336	9,054
Hospital D	1,234	65,636	92,960	3,371

cardiologists. Cardiologists' exposures average twice the radiologists exposures. Table 21 identifies some of the workload characteristics of the four hospitals included in Table 20. This information is provided so that realistic comparison can be made between the data reported for these Houston hospitals and one's own radiological facility.

7.3.1. Patient Exposure and Dose in Diagnostic Procedures

It has only been within the last 10 to 15 years that the primary concern of radiation control specialists has turned from that for radiological personnel to that for the patient. Two reasons account for this change in attitude. First, as has been shown, exposures to radiological personnel have been decreasing steadily and acute radiation-induced injuries have essentially disappeared. Secondly, although radiation exposures to patients have also followed a systematic reduction, the medical application of x-rays have increased significantly. One result of the 1964 U. S. Public Health Service population exposure study[308] was that an estimated 58% of the United States population had one or more x-ray visits during that year. Estimates for the continuing increase in medical x-ray utilization vary from 1% to 10% per year.

The recent literature is filled with studies of patient exposure and dose during various diagnostic procedures. For the interested radiologist, these studies present rather confusing information because the nature of the radiation exposure makes it difficult to identify a critical tissue and ascribe a precise dose to that tissue. Confusion is also generated because there is no consensus on what radiation units to employ in these instances. The three anatomical sites most often identified in diagnostic patient dose studies are the skin, the bone marrow, and the gonads.

7.3.2. Radiation Exposure of the Skin

The skin is often identified for measurement because the measurement can be made directly. Measurements of exposure to the skin have classically been made with ionization chambers or film. More recently the use of thermoluminescence dosimeters is gaining acceptance.[380,381] The results of measurements of radiation incident on the skin are variously recorded as exposure in Roentgens, or dose in rads.[382] For medical x-rays there is quantitatively little difference in these units. However, to confuse the issue further, many authors have used the units of integral dose (gram-rads), R-cm^2, and at least one[383] has reported an extensive series of measurements in terms of millijoules per examination. Although the millijoule is a precise physcial unit of energy, its use should not be encouraged because it conveys little useful information to the radiologist who is unfamiliar with the term. Further, suspected biological responses are classically related to the dose in rads.

Scintillation detectors[384] and commercially available transit ionization chambers are gaining wider acceptance as methods of monitoring patient exposure.[383,385] The transit ionization chamber is attached to the patient side of the collimators and intercepts the entire useful beam regardless of collimator position. The target to patient distance is unimportant because as the target to patient distance increases, the area exposed increases according to the inverse square law; likewise, with increasing target to patient distance, the radiation intensity decreases according to the same law. Consequently, these two factors cancel one another. The unit of measure with this type of an instrument is generally the R-cm^2[384-386] which has been identified as the dose-field product[387] or the surface integral exposure[388] (SIE). Patient doses measured for a number of different diagnostic procedures were reported by Carlson[389] as both integral doses in kilogramrads and in R-cm^2. Ardran et al.[390] measured skin exposure during cardiac catheterization with a transmission type ionization chamber and reported an average value of 1560 R-cm^2 from 54 patients. The average entrance exposure was 21.4 R.

Cameron[391] endorses this type of monitoring system for all routine diagnostic procedures. He notes that the available instrumentation is relatively inexpensive and that it would make possible the indentification of patient dose as an integral part of the patient's radiographic report. He objects, however, to the unit R-cm^2 and proposes in its stead the Roentgen area product (Rap). Cameron suggests that 1 Rap be defined as 100 R-cm^2. Upon initial consideration, this reviewer finds the introduction of yet another radiation unit unnecessary and undesirable.

Diagnostic film dosimetry has also been shown to be an effective means of estimating patient exposure.[392,393] Rogers[394] has published some rather extensive tables that show the skin dose for nearly all radiographic projections utilizing conventional radiographic technique. Some of these dose data are shown in Table 22 and the reader is referred to the original article for projections not included. The values in this table were calculated.

7.3.3. Bone Marrow Dose

Bone marrow dose cannot be measured directly, but this information may be of considerable importance in estimating a biological effect. The most sensitive latent somatic effect is apparently leukemia and its production is implied in the absorption of radiation by bone marrow stem cells. Similarly, the most sensitive short-term manifestation of radiation exposure is chromosome damage in circulating lymphocytes which is apparently a function of radiation energy absorbed at the site of lymphocyte production. Bone marrow dose may be estimated by various indirect ways.

Gough et al.[395] calibrated the output of diagnostic tubes in a cardiac catheterization laboratory so that patient exposure and dose could be estimated from predetermined operating parameters. They estimated the exposure to the skin for various procedures and from this data used the methodology developed by the Adrian Committee[396] to determine the bone marrow dose delivered during these procedures. They applied this methodology to 85 patients and published some rather extensive tables showing an analysis of each procedure. For example, 6 patients were fluoroscoped during pacemaker

TABLE 22

Examples of Calculated Skin Doses for Various Radiographic Projections Utilizing the Exposure Factors

	kVp	mAs	TFD in.	TFD cm	F	S	Skin Dose (mrads)
Skull							
Occipitofrontal	70	30	36	90	R	F	200
30° Frontooccipital (Townes')	75	50	36	90	R	F	370
Lateral	60	30	36	90	R	F	110
Submentovertical	80	50	36	90	R	F	470
Optic foramina	70	30	36	90	R	F	180
Optic foramina	65	25	28	70	R	F	120
Oblique (Stenver's)	85	60	28	70	R	H	1,010
Oblique lateral	85	70	28	70	R	H	1,180
Oblique lateral	85	100	36	90	R	H	880
Hand							
Posteroanterior	60	3	36	90	R	H	8
Lateral	60	30	30	75	I	–	120
Elbow							
Anteroposterior and lateral	60	10	36	90	R	H	31
Humerus							
Anteroposterior and lateral	70	50	30	75	I	–	340
Anteroposterior and lateral	70	10	36	90	R	H	47
Foot							
Dorsiplantar	80	4	36	90	R	H	23
Lateral	80	30	30	75	I	–	280
Knee							
Anteroposterior	85	60	30	75	I	–	700
Anteroposterior	70	10	36	90	R	H	49
Lateral	80	60	30	75	I	–	600
Lateral	65	10	36	90	R	H	38
Femur (lower 2/3)							
Anteroposterior and Lateral	70	5	36	90	R	F	26
Pelvis							
Anteroposterior	70	50	36	90	R	F	330
Anteroposterior	70	60	36	90	R	F	390
Lateral	80	60	36	90	R	F	840
Lateral	90	120	36	90	R	F	2,140
Cervical Vertebrae							
Anteroposterior 2–7	60	25	36	90	R	F	88
Lateral 1–7	70	20	60	150	R	F	43
Thoracic							
Anteroposterior	85	25	36	90	R	F	250
Lateral	75	40	36	90	R	F	490
Lungs							
Posteroanterior	65	8	60	150	R	F	13
Lateral	75	10	60	150	R	F	26

	kVp	mAs	TFd in.	TFd cm	F	S	Skin Dose (mrads)
Abdomen							
Anteroposterior	70	30	36	90	R	F	190
Stomach							
Posteroanterior	85	10	36	90	R	F	100
Posteroanterior	90	20	36	90	R	F	230
Enema							
Anteroposterior/posteroanterior	90	35	36	90	R	F	400
Lateral	90	60	36	90	R	F	1,130
Oblique	90	45	36	90	R	F	640
Pyelography							
Anteroposterior	70	30	36	90	R	F	200
Lateral	90	30	36	90	R	F	470
Pregnancy, at term							
Anteroposterior	65	60	36	90	R	F	420
Pelvimetry							
Inlet	90	125	36	90	R	F	1,680
Outlet	70	60	36	90	R	F	640
Lateral, erect	90	160	36	90	R	F	2,990

From Rogers.[394]
Legend:
TFD: target to film distance
F: type of film
R: Red seal
I: Ilfex
S: type of screen
F: fast
H: high definition

insertions and the average skin dose was estimated to be 132 rads per patient, while the average bone marrow dose was 3.6 rads.

Liuzzi et al.[397] employed film monitoring to estimate the average bone marrow dose during fluoroscopy. They subdivided the incident exposure field of the trunk and totaled the dose contribution to each of these subfields using large monitor films attached to the fluoroscopic table. This allowed them to calculate volume weighted average bone marrow doses for each subfield. Their methodology was explained thoroughly, but there was little clinical information in their report. Yoshinaga et al.[398] extended Liuzzi's methodology to generate clinically useful data. However, they attached the monitoring film to the subject rather than to the x-ray table. The average bone marrow dose during upper G. I. examinations was 105 gram-rads for the fluoroscopic portion and 72.6 gram-rads due to spot filming. These values were based on ten patients.

The Adrian Committee[396] in 1966 published a rather complete analysis of bone marrow doses for various radiological procedures in Great Britain. The results were based on a population x-ray survey conducted in 1957-1958. This analysis assumed that irradiation of any part on the whole of the active bone marrow should be averaged over the whole bone marrow. Consequently, the resulting doses are not simply those to the bone marrow contained within the primary beam. This investigation showed that the bone marrow dose from general diagnostic radiology per person averaged over the entire population was 24.4 mrads per year. When mass miniature radiography, dental radiography, and the diagnostic use of radioisotopes were included in the analysis, the result was 32.4 mrads per year. The average bone marrow dose per diagnostic radiological examination was about 100 mrads. Table 23 contains data abstracted from this report on the average bone marrow doses experience during various proce-

TABLE 23

Average Bone Marrow Doses Resulting From Various Diagnostic X-ray Examinations in National Health Service Hospitals, Great Britain

	Average Marrow Dose Per Examination (mrads)		Average Annual Marrow Dose Per Person (mrads)		
	Male	Female	Male	Female	Total
Barium meal	505	799	2.39	3.58	5.97
Barium enema	528	1060	1.32	2.04	3.36
Whole chest	12	13	1.01	0.83	1.84
Lumbar spine	270	270	2.05	0.98	3.03
IVP	584	452	1.37	0.92	2.29
Abdomen	123	129	0.38	0.39	0.77
Abdomen, Obstetric		206	----	0.33	
	500*		0.40*	0.40*	1.13
Pelvimetry		283	----	0.11	
	1120*		0.21*	0.21*	0.53
All other examinations	----	---	1.92	1.64	3.56
Private practice examinations					1.9
					24.4

* dose to fetus

From the Adrian Committee.[396]

dures. The population estimates contained in this table take into account the absolute frequencies of each type of examination by sex.

7.3.4. Local Tissue Dose During Mammography

Although radiographic mammography had been employed for many years, it was not until the work of Egan[399] that it received widescale acceptance among general radiologists. Mammography, utilizing the Egan technique, is contrary to the trend of reducing patient dose during diagnostic procedures, through the use of high speed film-screen combinations, low mA, and high kVp technique. Modern mammography utilizes low kVp and filtration, high tube current and time, and fine grain, slow speed film. These conditions result in a considerably higher radiation exposure than most other radiographic procedures and which is of special concern because of the increasing utilization of mammography. Some estimates of mammographic radiation dose are based on phantom measurements and calculations.[400-402] Others have utilized thermoluminescence dosimeters affixed to various anatomical patient locations.[403-405] In the series reported by Gilbertson et al.[403,404] radiation exposures varied from background at distal portions of the anatomy to 25 R to the surface of the inner breast when the x-ray tube had an inherent filtration of 0.4 mmAl. When this filtration was supplemented with an additional 0.38 mmAl, the exposure to the surfacc of the inner breast averaged 18 R. When the two axillary views were omitted from the examination, as might be proposed for any large population survey, the exposure to the surface of the inner breast averaged 11 R for the higher filtration technique. In a preliminary study reported by Palmer et al.[405] the average dose during a 6 view mammographic exam of 10 patients varied from 6.4 rads to 21.6 rads to the breast depending upon where the thermoluminescence dosimeters were placed.

7.3.5. Gonadal Exposures

Many reports have appeared during the past ten years that deal with exposure of the gonads during diagnostic procedures. Although direct measurement is most often employed in these studies, computation methodology has also been reported. This methodology relies on a knowledge of x-ray operating parameters and is particularly suited for large-scale surveys. Table 24 contains information developed by Heller et al.[406] using their computational methodology to estimate gonadal doses

TABLE 24

Average Adult Gonadal Doses (in mrads) for Various Radiographic Procedures as Performed by New York City Physicians

Examination	Male	Female
Chest	0.38	1.30
Hip	4020	855
Lumbar spine		1188
Pelvis	1999	437
Retropyelogram	2635	1130
Pelvimetry		485

From Heller et al.[406]

resulting from examinations in physicians' offices in New York City.

Table 25 contains data extracted from Matthews and Miller.[407] The gonadal dose values contained in this table represent measurements made over an 18-month period in 1964-65 on 878 patients receiving 2184 exposures during the various radiological examinations. The measurements were made at 30 different radiological facilities.

Table 26 is presented to show the relative inconsistency in measurement for one type of high gonadal dose examination—intravenous pyelography. The measurements span an 18-year period and represent studies from several different countries. The inconsistency of measured doses described by this table are evident also for other types of radiological examinations and are apparently due to differences in equipment and techniques.

Because of the possible long-term genetic significance, particular attention has been paid to gonadal doses during gynecological examinations and gonadal doses to children.[414-417] Niki[414,415] utilized an inhomogeneous tissue equivalent phantom to measure ovarian radiation dose during routine gynecological radiographic and fluoroscopic examinations. He presents considerable data on dose to the ovaries depending upon ovarian position, type of exposure, and exposure factors. The exposures ranged from a low of 28 mR per examination to a high of 496 mR. Aspin[416] has reported gonadal doses in a child simulating phantom measured with ionization chambers for a number of different types of examinations. Kaude et al.[417] reported gonadal doses in children during voiding cystography with thermoluminescence dosimeters. The examination was performed with 70 mm fluorography under video fluoroscopic control and the measurements were made on 41 patients. The average testicular dose was reported to be 105 mrads and the average ovarian dose was 269 mrads.

Some papers have also appeared[418,419] reporting gonadal doses following radiotherapeutic treatments. The studies were undertaken because as the cure rate from some therapeutic procedures increases, the likelihood of progeny from the patients so treated also increases. Gonadal doses received by these persons are then of some concern from the genetic standpoint. Gonadal doses in the hundreds of rads are not uncommon during some types of treatments.

TABLE 25

Average Gonadal Exposures for Various Types of Examinations

Type of Examination	Male mR/exam	Female mR/exam
Head and neck	0.24	0.49
Arm, hand	0.35	0.12
Shoulder	0.06	0.28
Chest	0.88	0.21
Barium meal	5.4	201
Abdomen	31.2	183
Intravenous pyelogram	458	312
Barium enema	95.0	692
Pelvis, lumbar spine, lumbar-sacral joint	148	471
Hip, upper femur	1,216	252
Leg, foot	0.42	11.2

From Mathews and Miller.[407]

TABLE 26

Representative Examples of Gonadal Dose During Intravenous Pyelography as Determined by Various Investigators

Investigation	Date	Gonadal Dose per Examination (mrad) Male	Female	Ref.
Stanford and Vance	1955	486	1290	408
Hammer-Jacobson	1957	1381	424	409
Laughlin-Pullman	1957	100-2000	200-1200	410
Larsson	1958	1240	925	411
Adrian Committee	1960	804	637	412
McEwan	1966	884	355	413
Mathews and Miller	1969	458	312	407

7.3.6. Genetically Significant Dose

Brecher[2] states that the reduction of dose to patients during routine radiological examinations from 1942 to 1966 was approximately tenfold. For this reason, measurements of skin and bone marrow doses are of less importance today than they were then. The measurement of gonadal dose, however, is assuming increasing importance because of the expansion of techniques and the increase in frequency of diagnostic radiological examinations.[308,420,421] The previous section discussed the methods and extent of investigation designed to determine gonadal dose during radiological procedures. Using this type of information, and additional information relative to frequency of x-ray examinations on an age-specific basis, estimates of the genetically significant dose (GSD) have been made.

The genetically significant dose is defined as the gonad dose which if received by every member of the population would result in the same total genetic effect to the population as the doses which are actually received by individuals within the population.[422]

In other words, the genetically significant dose (GSD) is a measure of the total radiation received by the genetic population pool, which may have a possible effect on the progeny of that population. The determination of the numerical value of the GSD is adjusted to allow for the expected progeny of individuals within the population and it depends upon: 1) the average gonadal dose from various examinations for patients in a given age-sex category, 2) the number of persons in each age-sex category who received a specific type of radiologic examination during the year, 3) the number of persons in each age-sex category in the total population, and 4) the expected number of future children for persons in each age-sex category.

Estimates of the GSD have been made more frequently and for more population groups during the last 10 years than for the prior 65 years combined. The concern for radiation induced genetic effects is certain to continue, and, therefore, no doubt, so will the frequency of these GSD estimates.

In a 1967 report, Mahmoud et al.[423] determined the average annual GSD due to diagnostic x-ray exposure during the period 1955 to 1961 for two cities. The results showed a GSD in Alexandria, Egypt, of 7.2 mrad per year while the GSD for a large section of the city of Cairo, Egypt, was 28.7 mrad per year. The differences in GSD were explained on the basis of differences in population disease and examination practices.

The GSD was reported to be 75.3 mrads in New Orleans for the year ending March 31, 1963.[424] It was noted that approximately one-third of this GSD was due to obstetrical examinations. The GSD in New York City during the year 1962 was estimated to be 50 mrads.[425] Approximately 34 mrads were contributed by general radiographic procedures while 16 mrads were due to fluoroscopy.

Two of the most authoritative studies dealing with the GSD are those of the Adrian Committee[396,412,426] and the 1964 studies conducted by the U. S. Public Health Service.[308,421,422,427] These reports have presented rather complete analyses of radiological examination frequency, estimated gonadal dose perexamination, and contributions to the GSD as a function of age and sex for various examination categories, including type of medical practice and type of radiological facility.

7.3.7. Adrian Committee Reports

The Adrian Committee was organized in 1956 by the Ministry of Health for Scotland to review the practice of diagnostic radiology and its contribution to population radiation dose. National surveys were conducted during 1957 and 1958 from which their statistical data relating to examination frequency were obtained. Physcial measurements were obtained at the time of the national survey and subsequently were used to estimate gonadal dose. The calculations were based on an estimated 13 million x-ray examinations in Great Britain during 1957. The Committee's results indicated a GSD at that time of 14 mrads per year from diagnostic x-ray exposure.

7.3.8. 1964 Public Health Service Survey

The U. S. Public Health Service survey of 1964 was considerably more extensive than the Adrian Committee survey. The Public Health Service study was intended to provide accurate estimates of exposure and dose for the United States population based on the x-ray experience of a large representative sample. The study was conducted in several sequential stages:[422]

1. a household interview survey of a representative sample of the U. S. population to obtain x-ray visit data.

2. a mail follow-up of the x-ray facilities where reported visits took place to obtain technical data.

3. the merging and processing of the above sets of data to produce estimates of the x-ray experience of the U. S. population.

4. the development of dosimetric models for estimating gonad dose from the exposure data generated by the x-ray exposure study.

5. the estimation of gonad doses and the genetically significant dose for the U. S. population based on the dosimetric models and the exposure data.

The exposure study[308] resulted in an estimate that during 1964, 180 million persons, or 58%, of the civilian population of the United States had one or more x-ray visits. Approximately 66 million persons had radiographic examinations, 46 million had dental x-ray examinations, and 8 million had fluoroscopic examinations. The medical radiographic examinations accounted for nearly all of the genetically significant dose from diagnostic x-rays and were responsible for essentially all of the unnecessary gonadal radiation exposure. Eighty-two percent of the estimated GSD resulted from examinations of males, while only 16% resulted from examinations of females. The contributions to the GSD from fetal exposure amounted to only 2%. Radiography contributed 96% to the GSD, fluoroscopy 4%, and less than 1% from photofluorography. Males between the ages of 15 to 29 years were the largest single population group contributing to the GSD.

Table 27 shows the estimated percent of distribution of medical x-ray procedures by type of facility. Nearly 60% of all x-ray procedures were conducted in hospitals and 60.8% of all procedures were supervised by radiologists. Those procedures not supervised by radiologists were approximately equally distributed among internists, general practitioners, orthopedic surgeons, and others. Table 28 reports the relative frequency of radiographic examinations for the various body areas and the contribution that examinations of these areas made to the GSD. Fifty-one percent of all radiographic examinations involved the thoracic area, and yet these examinations contributed only 4% of the GSD. On the other hand, lower abdominal examinations constituted only 14% of all radiographic examinations, and yet they contributed greater than 80% of the GSD. Table 29 shows the results of analyses of fluoroscopic examinations by type of examination and contribution to the GSD. The total contribution to the GSD from fluroscopy was approximately 4% and most of this was due to barium enema examinations. Gastrointestinal examinations constituted 56% of the total fluoroscopic exams, but contributed only 1.3% of the GSD.

TABLE 27

The Estimated Percent Distribution of Medical X–ray Procedures by Type of Facility. 60.8% of all procedures were supervised by radiologists.

Type of Facility	Percent Medical X–ray Procedure
Hospitals	57.6
Private Radiologists Offices	4.5
Other Private Offices	15.8
Private Groups	6.6
Health Agencies and Others	15.5

From Gitlin and Lawrence.[308]

TABLE 28

The Estimated Percent Distribution of Radiographic Examinations by Body Area and the Contribution of these Examinations to the GSD

Body Area	Percent of Examinations	Contribution to GSD mrads	Contribution to GSD %
Thoracic	50.6	2.3	4.2
Lower abdominal	13.7	46.0*	84.2
Upper abdominal	11.0	2.4*	4.4
Lower extremities	9.7	2.6	4.8
Upper extremities	7.3	0.1	0.2
Head and neck	7.1	0.1	0.0
Multiple areas	0.6	1.2	2.2
	100.0	54.6	100.0

*Includes fluoroscopic contribution. From Gitlin and Lawrence[308] and Brown et al.[422]

7.3.9. Summary

Determinations of the GSD for a number of population groups throughout the world show that although the United States does not have the lowest estimated GSD, neither does it have the highest. Table 30 is a comparison of the published

TABLE 29

The Estimated Percent Distribution of Fluoroscopic Examinations by Type of Examination and their Relative Contributions to the GSD

Type of Examination	Percent of Total Examinations	Contribution to the GSD mrads	percent
Gastrointestinal series	56.4	0.7	1.3
Barium enema	29.0	1.5	2.8
Other	14.6	0.1	0.2
	100.0	2.3	4.3

From Gitlin and Lawrence[308] and Brown et al.[422]

TABLE 30

The Estimated GSD Due to Diagnostic Examination for 16 Different Population Groups

Population Identification	Time of Study	GSD mrads	Ref.
Alexandria, Egypt	1955-61	7	423
New Zealand	1963	12	428
Great Britain	1957-58	14	412
Texas	1960-62	16	429
Johns Hopkins University	1963-64	20	383
Denmark	1956	22	409
Cairo, Egypt	1955-61	29	423
Japan	1960	39	430
Richland, Washington	1953-56	43	431
Rome, Italy	1958	44	432
Kaiser Hospital	1956-57	40-50	433
New York City	1962	50	425
United States	1964	55	422
Sweden	1955	72	411
New Orleans	1962-63	75	424
Australia	1950-55	159	434

values of the GSD for 16 different population groups. The values range from 7 mrads per year for Alexandria, Egypt, to 159 mrads per year for Australia. The United States level of 55 mrads is slightly greater than the numerical average of 44 mrads per year. Although these values for GSD are only estimates, several are from rather large and scientifically precise investigations.

This review has not considered the human evidence for radiation induced genetic effects because as Westing[435] pointed out 30 years ago there is none available in the dose range of diagnostic x-rays. A GSD of 50 to 100 mrads per year is obviously an insignificant and almost immeasurable quanity of radiation in light of our knowledge of radiation induced somatic effects on man. It is in the range of naturally occurring background radiation. However, because the long-term genetic effects are unknown, we must continue to be prudent in our medical application of x-rays so that the GSD remains at this low level.

8. EXPOSURE LIMITING DEVICES AND PROCEDURES

The primary aim of modern diagnostic radiology is to obtain the maximum medical information while subjecting the patient and the operating personnel to a minimum radiation risk. Currently employed radiological equipment and techniques result in radiation exposures to these persons that are orders of magnitude below exposures experienced just a few years ago. The continuing reduction of these exposures is a function of a number of aspects of contemporary diagnostic radiology. Among these are educational opportunities and requirements, equipment and accessory development, procedural and managerial innovations, and careful attention to individual techniques.

Many of the developments in diagnostic radiology in the past that have served to reduce individual radiation exposure have been rather easy to identify and define. Such devices as grids, screens, diaphragms, image intensifiers, and faster films are readily recognized as contributing to lower radiation exposures. Equipment developments such as these no doubt will continue in the future and will help support the downward trend in diagnostic radiation exposure while sacrificing nothing in medical information. However, the more subtle aspects of diagnostic radiology are becoming increasingly important to radiation exposure reduction. More careful training of all radiological personnel, better managerial efforts in diagnostic departments, and more selective judgment in the performance of a particularly necessary or unnecessary examination are areas in diagnostic radiological practice of importance today and promise to be even more critical in the future. Some of these areas were discussed at length in the 1967 report of the U. S. Public Health Service Medical X-Ray Advisory Committee.[436]

8.1.1. Equipment and Accessory Design

Some of the accessory devices that are taken for granted today were rather innovative developments in their time. Generally, the devices that contribute to increased radiological information also are effective in reducing radiation exposure. The purpose of this review will be served by taking note of these developments. Table 31 contains a summarization of our more important equipment standards as obtained from the most recent reports of the various radiation advisory groups.

8.1.2. Filtration

Brecher[2] states that within a year following Roentgen's discovery of x-rays, filters were shown to be effective in producing sharper radiographs. Although aluminum filters are used now, for many years leather was the filtration of choice. The filter in an x-ray machine serves the purpose of differentially absorbing the low energy x-ray photons. Were these photons not filtered from the primary beam, they would result in little significant radiological information because they are readily absorbed by the patient and, consequently, they would serve only to increase the patient dose. The required amount of filtration for producing maximum diagnostic information depends upon: a) the kVp, b) the anatomical location and thickness of the body, c) the type of examination, and d) the film and screen characteristics employed. For fluoroscopy a minimum total filtration of 2.5 mmAl is required. For radiographic application, the same total filtration is required for operation above 70 kVp. For operation below 50 kVp, 0.5 mmAl is necessary and for intermediate tube potentials 1.5 mmAl is required.

This level of filtration is high compared to that employed earlier. During the 1920's and earlier, total filtration exceeding 0.5 mm of aluminum was rarely used. DeLorimier et al.[437] have described the protective features in an army field x-ray unit used during World War II. For radiography, a 0.25 mm aluminum filter was added, which with the inherent filtration resulted in a total filtration of 0.5 mmAl. During fluoroscopy, an additional 0.5 mm aluminum filter was used. Perhaps the studies of Trout et al.[438,439] have been most instrumental in increasing the required filtration for radiological equipment. Their studies have shown that there is an optimum filtration at each operating potential that will result in maximum radiological information and, at the same time, minimize radiation exposure to the patient. They have urged equipment manufacturers to include devices that would change filters automatically with changing kVp and perhaps this will come to pass with future equipment.

8.1.3. X-Ray Beam Alignment and Collimation

One of the major deficiencies in modern radiographic equipment is that beam collimation and alignment devices do not adequately restrict the useful x-ray beam to the area of the fluorescent screen or film, much less to that part of the body being examined. Two factors are involved. First, the x-ray beam should be aligned so that its central axis can be easily positioned to intercept the

TABLE 31

Some Current Standards for Diagnostic Radiological Equipment and Accessories

Item		Standard
1. Protective x–ray tube housing		<100 mR/hr leakage radiation at one meter
2. Fluoroscopic target to table top distance		>12 inches (30 cm) and preferably 15 inches (38 cm)
3. Total filtration		≥2.5 mm Al
4. Scatter radiation shield on fluoroscope		≥0.25 mm Pb barrier
5. Reset timer		5 minutes allowed before reset
6. Protective apparel		at least 0.25 mm Pb equivalent
7. Beam restricting devices		<100 mR/hr leakage radiation
8. Adjustable collimation		light localizer and x–ray beam accurate to within 1 inch at 72 inch distance.
9. Total filtration	<50 kVp	0.5 mm Al
	50-70 kVp	1.5 mm Al
	>70 kVp	2.5 mm Al
10. Beryllium window tube		must have keyed filter interlock

center of the film or screen through the center of the area under examination. Secondly, the use of cones and collimators, or diaphragms, to limit the dimension of the beam must be properly designed and adjusted. These rudimentary features of radiation control were probably first reported by Rollins, the Boston dentist, in 1901.[440] The Medical X-Ray Advisory Committee[436] has identified the major causes of poor collimation and alignment: a) cones, suitable for all required field sizes, are not usually available; b) when correct cone size is available, larger cones are often used anyway because of the difficulty in aiming the x-ray beam; c) often cones are not accurately calibrated and do not fit properly on the tube head; d) the geometrical principles involved in determining projected field sizes for varying source to film distances are difficult for many operators.

Variable aperture collimators with built-in light localizing devices are helping to solve many of these difficulties. However, even this type of beam defining apparatus is easily misadjusted so that the light field and the x-ray field do not coincide properly. Beam alignment standards for modern x-ray equipment require that the field sizes indicated on the variable collimators must be accurate for both the x-ray field and the light field to within one inch for a source of film distance of 72 inches. The requirements based on fixed aperture cones are not quite so strenuous and they allow for a 2-inch margin at a source to film distance of 72 inches, or a 1-inch margin at a source to film distance of 36 inches.

Proper beam alignment may appear at first glance to be a simple matter. However, considerable attention is being given to this one aspect of radiological equipment because of apparent widespread inadherence to the recommendations. One major finding of the late 1964 population dose study[422] was the estimation that if the primary x-ray beam were restricted to an area no larger than the film size on all x-ray machines, then the GSD would be reduced from 55 mrads to 19 mrads per person per year.

Checking beam alignment on an x-ray machine is not difficult. Hale[441] has described the effect of improper collimation on patient dose and on the diagnostic information obtained. He suggests methods of optimizing both of the factors to the benefit of examiner and patient. The most straightforward way in a private office is to expose single sheet film wrapped in paper to an x-ray beam, after having first affixed conspicuous markings to the film showing where the x-ray beam should be on the basis of a light localizer or geometrical considerations. Other devices which are now commercially available utilize a crossarm strip of fluorescent material that has scale factors and on which one can visualize fluorescence produced by the x-ray beam in a dimly lit room.[442] The proper identification and alignment of radiographic cones is not so simple. Blatz and Eure have described a "cone calibration computer" to aid in the proper selection and calibration of radiographic cones.[443]

8.1.4. Supplementary Patient Shielding and Positioning Devices

One of the greatest excesses in diagnostic radiology is the number of repeat radiographs that are made. Repeat radiographs result from a number of things, including improper technician training, incorrect patient positioning or patient motion, and improper film development. Proper attention should be given to each of these factors because a considerable saving in time and money is possible, not to mention the reduced patient dose. When the patient, or the examination, requires it, mechanical restraining devices should be used. The patient should not be held by another individual unless it is absolutely necessary, and then the person assisting should not be a radiation worker and should be provided with proper protective apparel.

The shielding device of primary importance to patients if the gonadal shield. Gonadal shields should be employed during all examinations in which the gonads are in the primary beam and when they will not interfere with the desired radiographic information. It should be kept in mind, however, that the gonadal dose reduction afforded by a protective lead shield may be considerably less than expected, because a great percentage of radiation reaching the gonads will have been scattered axially through the body to the gonads.[419,441] The recent radiological literature abounds with descriptions of gonadal shielding devices,[411,444-456] but as Steadman[457] points out, the availability of gonadal shields in a diagnostic x-ray department does not mean that these shields will be used. In fact, many would agree available gonadal shields are used only a fraction of the time that they should be. One of the most interesting and novel approaches to gonadal shielding has been proposed by Smith and Willhoit.[458] They have studied the shielding properties of a lead line foundation garment designed for use by female radiation workers. They determined gonadal dose reduction factors up to 19, depending upon radiation quality and angle of incidence. However, it seems unlikely that this device will ever gain wide acceptance.

In addition to the gonads, the lenses of the eyes are considered to be particularly sensitive to radiation. During tomographic procedures it is not uncommon to have radiation doses exceeding 10 rads delivered to these organs.[459] Simple lead shielding devices have been described for this radiographic situation.[460]

8.1.5. Diagnostic X-Ray Imaging Devices

The X-radiation that exits from a patient's body during a diagnostic examination contains the useful radiological information. The manner in which these x-rays are received and the information portrayed contribute greatly to the patient dose. The imaging devices include film-screens, films, fluoroscopic screens, electronic image intensifiers, and other methods. The influence that these accessories have on patient dose is well known and has been rather thoroughly discussed in Janower's book[461] and these accessories were more recently considered at the 13th Annual Meeting of the American Association of Physicists in Medicine.[462]

Grids are used to reduce the amount of scatter radiation, generated in the patient, that reaches the x-ray film. They are usually built into the film cassette and different types of grids are used, depending upon the machine operating parameters, the size of the radiation field, and the thickness of the patient. The use of grids generally requires an increased exposure because they absorb some of the primary beam radiation. But the gain in diagnostic information more than offsets the risk involved with the higher radiation exposure.

Intensifying screen-film combinations require only approximately 5% of the radiation that film alone requires to produce the same image. Undoubtedly the great majority of x-ray films taken today employ double emulsion film contained in cassettes with intensifying screens both front and back. There are several categories of intensifying screens and their use must be tempered by an understanding of the quality of information desired. Generally speaking, the higher the screen speed, the lower the patient radiation dose, but also the lower the image resolution. Consequently, compromises are most often necessary between speed and image definition.

The comments relative to intensifying screens are applicable to radiographic film also. The physical factors of radiographic film affecting radiographic image quality have been extenisvely studied and reviewed by Bates.[463] Maximum diagnostic information, coupled with minimum patient dose, is obtained by the proper choice of film with reference to speed, contrast, and latitude, and by attention to consistent processing technique. Goldberg[464] has reported the development of a simple device that would obviate retakes because an unloaded cassette was used. He claims that adoption of this device has significantly reduced the number of retakes in his facility.

The development of fluoroscopic imaging from the intensifying screen through the electronic image intensifier to stop action type electronic devices is an interesting evolution that has been accompanied by significant reductions in patient dose while increasing the quality and quanity of radiological information. Properzio and Trout[465] have shown an approximate tenfold reduction in patient exposure over a 40-year period, due to improvements in the luminescent response of commercially available fluoroscopic screens. The use of an image intensifying tube in fluoroscopy can easily result in an additional tenfold decrease in patient exposure. The development of stop action fluoroscopy based on the use of intermittent x-radiation with image storage and video playback is in the early stages of development but offers promise of significant improvement in diagnostic radiology.[466-468] At the present time, the available instrumentation is quite expensive and the image quality not as good as one would like. Docker and Astley[466] report encouraging results of retained image quality with the dose reduction factor of 15 to 20.

8.1.6. Protective Apparel for Radiological Workers

The use of leaded gloves and aprons is well established in diagnostic radiology. This has not always been the case; in fact, it was not until perhaps the early 1950's that these devices received widespread general acceptance. Modern aprons and gloves come in many sizes and thicknesses—the most common thicknesses being 0.25 mm, and 0.50 mm, and 1.0 mm of lead equivalent. Protective apparel of 0.25 mm lead equivalent satisfies the minimum recommendations. However, apparel of 0.5 mm lead equivalent is more desirable beacuse of the added dose reduction factor. Granted, the more protective apparel is heavier, but it is the experience of most radiologists that a 0.5 mm lead equivalent apron can be comfortably accommodated.

Some early experiences with protective apparel have been described by Kaye[55] and White et al.[469] It is apparent from these early discussions that acceptance of leaded apparel by radiologists was not easy. One rationalization for not wearing protective apparel as given by Cilley et al.[58] was that during examination of the barium filled stomach, the barium itself provided adequate shielding for the ungloved hand when positioned in the middle of the radiation field. Even today many radiologists glove only one hand, or perhaps neither hand during fluoroscopic examinations, even though some recent data show that these ungloved extremities[287] may be receiving a restrictive dose of radiation. Occasional innovations and alterations to our conventional protective apparel appear. Shielding designs other than for gloves have been described for the protection of the upper extremities during fluoroscopic examinations.[470] Ross has described a detachable extension to a conventional protective apron that provides shielding for a greater portion of the lower extremities.[471]

8.1.7. Special Examinations

Mammography has been described as a high local dose procedure because of the low filtration and high tube current employed. Two papers[472,473] have shown that significant local tissue dose reduction can be effected with little loss of radiographic information by filtration designs not normally used. Hranitsky and Shalek[472] showed that additional aluminum filtration could be used during mammography with resulting skin dose reductions from 23 to 2 rads per film. When Maudal[473] employed a combination of aluminum and iron filtration, the skin exposure per film was reduced from 24.5 R to 4.2 R. Price and Butler[474] described a significant dose reduction during mammography by using an x-ray tube with a rotating molybdenum anode. This type of tube when used in conjunction with a screen film combination produced acceptable mammograms with an exposure time of only 0.4 seconds and an accompanying small radiation dose.

Cinefluorography is a technique that produces serial still images in sufficient number to provide the cine effect. It is also a high dose procedure with skin exposures in the order of 5 to 10 R for a 10 to 15 second sequence of 16 frames per second. Research and development on equipment is continuing in this area of diagnostic radiology and radiologists are becoming more cautious by reserving the use of cine for situations truly indicated. Likewise, general fluoroscopic examinations are high dose procedures and should be reserved for clearly indicated situations. Fluoroscopy should never be a substitute for radiography. In addition to the previously enumerated developments in image recording devices, more flexible examination tables are being developed,[475] and various types of safety circuits are being installed as integral accessories to examining room equipment.[476,477]

8.1.8. Radiation Control Surveys of Diagnostic Equipment

Specifications for safe radiological equipment are widely published and generally accepted. Some of them now carry the force of law in the United States resulting from the passage of Public Law 90-602 entitled Radiation Control for Health and Safety Act of 1968. This law essentially requires x-ray manufacturers to meet federal standards in the manufacture of their radiographic equipment. For the most part, these federal standards are direct adoptions of the NCRP recommendations.

With the establishment of these regulations, the routine radiological survey of diagnostic facilities becomes more important. Radiological surveys and calibrations of radiotherapy equipment have been standard practice for many years. Only now is this also becoming widely accepted in diagnostic departments.

A number of suggestions have been submitted for the proper conduct of radiation protection surveys.[330,478,479] Basically, a thorough survey of diagnostic radiological equipment should consist of the following general observations or measurements: a) a radiation output for the various normal operating conditions, b) leakage radiation from tube housing, c) scatter radiation levels during fluoroscopy with a patient simulating phantom in the beam, d) protective apparel and devices (aprons, gloves, Bucky slot cover, fluorescent screen curtains, dead man fluoroscopic switch, cumulative reset timer, etc.), e) estimation of workload, f) proper beam alignment and collimation, g) adequate filtration, h) adequate primary and secondary shielding barriers, i) proper darkroom construction and adequate processing technique, j) equipment inventory, and k) personnel monitoring records.

In the few large surveys that have been conducted recently, some rather interesting findings have resulted. In the Polk County, Florida, survey, for example,[314] 51 out of 139 radiographic units required additional filtration. Of these units, 59 had improper beam size or alignment. In a survey of x-ray facilities in the Federal Health Program Service[316] the greatest deficiency note was that 75% of the mobile x-ray machines had a control switch cord that was too short. The study revealed that 11% of the medical machines and 32% of the dental machines were equipped with inaccurate and inadequate timers. Results of the radiation control surveys of x-ray facilities within the Bureau of Prisons System showed that 68% of the dental x-ray timers were inaccurate and 34% of the medical x-ray machines required additional filtration.[315] Since 21% of the x-ray machines did not have the proper cone available, a variable aperture collimator was recommended.

The results of the radiation control surveys of over 5000 medical x-ray facilities during 1962-1967 led Fess and Seabron[313] to speculate on the characteristics of the average medical diagnostic x-ray facility. Only one person out of every 2.5 persons in this hypothetical facility is provided with a personnel monitoring device. The x-ray tube housing has an 87% chance of meeting the standards for leakage radiation. The tube housing will have a 50% chance of having a total filtration of at least 2.5 mmAl. In the typical facility the radiographic workload is 47.4 patients per week, and in the fluoroscopic workload 11.3 patients per week. Gonadal shields are present only 38% of the time and, more often than not, are not employed. Each radiographic x-ray head has a 30% probability that it is improperly collimated by at least 2 inches or greater at the maximum film size. The fluoroscope in this facility will be equipped with shutters which properly limit the beam to the fluorescent screen in only 50% of the cases. The average user of the fluoroscope dark adapts approximately 16 minutes before use. Adequate Bucky slot covers and leaded drapes around the fluoroscopic screen have only a 50% and 30% chance, respectively, of being present. The average fluoroscopic examination time is 3.1 minutes. The probability that automatic film processing is employed is only 4%. Fluoroscopic image intensifiers are available on only 8% of the machines.

8.2.1. Radiological Procedures and Techniques

Adequate radiation control cannot be accomplished with even the best of equipment and devices unless proper procedures are employed. These procedures relate to actions which serve to reduce exposure to both operating personnel and patient and they are basically tied to common sense considerations. Enumeration of some of these procedures may be helpful.

Every diagnostic facility should be inspected and evaluated annually by a competent radiological physicist or health physicist who has had sufficient experience in diagnostic radiology. Prior to the modification of an existing examination room and before construction of any new facility, a competent physicist should be called upon to evaluate the total radiation control aspects of the proposed facility. At a preconstruction consultation, the physicist's primary responsibility would be to prescribe shielding requirements and size requirements for the examination rooms. He may then assist in the proper positioning of the radiographic equipment.

8.2.2. Radiological Personnel

The biggest problem in the protection of radiological personnel arises because familiarity with the surroundings and the machines tends to diminish one's respect for the radiation hazard. Most diagnostic departments do not have area radiation monitors; therefore, since the personnel cannot sense the radiation, they tend to forget that a radiation field may exist in their working environment. They tend also to take shortcuts in their daily assignments that often lead to unnecessary increased radiation exposure. The following rules are some that should be strictly adhered to by all diagnostic x-ray personnel.

1. During fluoroscopy proper protective apparel should be worn. This means an adequate apron and two protective gloves. During some special procedures it may not be possible to glove both hands, but every effort should be made to do so.

2. During fluoroscopy the tube energizing foot switch should be used as intermittently as possible. The examiner should not activate the switch and continue to stand on it.

3. Use the smalllest possible field size during each examination.

4. Only those persons absolutely necessary should be present in the examination room.

5. X-ray workers should not be employed to hold patients during radiographic examinations. Individuals who are obtained for this purpose should be provided with complete protective apparel and should be carefully instructed to remain as far from the useful beam as possible.

8.2.3. Patient Exposure

Common sense also plays an important role in adopting procedures to help limit exposure to the patient. Some have already been discussed, such as proper operation of the radiological equipment. Chamberlain[480] has discussed others and gives us a good example from the manner in which an upper extremity is radiographed. The patient should not be seated with his legs extending under the examining table. Rather, the patient should be turned so that his back is to the examining table and if it is a modern general purpose table, the built-in side shield should be used. If the side shield is not available, the patient should be protected with a protective lead apron.

Most radiologists and radiation scientists would agree that a major contribution to patient dose in diagnostic procedures is the fact that many unnecessary examinations are conducted. The radiologist is often caught between two conflicting views. The referring physician requests a certain particular radiological examination and the radiologist realizes that the examination is unnecessary and, therefore, he would prefer not to conduct the examination. If he does not, he may antagonize the referring physician; while if he does conduct the examination, he is not following what he recognizes as proper radiological practice. To carry the paradox further, it is possible that the radiologist could be the subject of litigation for refusing to conduct an examination later judged to be necessary. On the other hand, he might become the subject of a lawsuit for conducting an unnecessary examination that later was shown to contribute to some latent radiation effect. Nevertheless, it is prudent practice for the radiologist never to perform an x-ray examination unless there are sound medical indications. The radiologist should be reasonably aware of the patient's recent medical radiation history so that repeated examinations may be obviated. In other words, a radiologist would be fulfilling his responsibilities if he refused to refer a patient for a radiologic examination when in his judgment it was not medically indicated.

Some rather clear examples of diagnostic x-ray application that generally are not medically indicated are: a) chest x-ray for routine pediatric checkup, b) gastrointestinal series for a two-months pregnant woman suffering nausea, c) x-rays for differential diagnosis when pregnancy is suspected except when there is strong clinical evidence of malignancy, or other emergency conditions, d) x-ray examination of the lower abdomen or pelvic region during pregnancy, especially in the first trimester except for the circumstances just listed, e) full mouth x-ray examination for routine dental checkups.

Perhaps the most critical situation during which unnecessary radiation exposure occurs is during pregnancy. Routine chest examination during pregnancy should be discontinued. The low frequency of positive findings from this screening examination does not warrant the radiation hazard involved. Other types of examinations are even more questionable. Ovarian radiation and fetal radiation dose during pregnancy radiographs can amount to several rads. However, using a high kVp and high filtration technique, these doses can be reduced as much as 85%.[481] The earlier in pregnancy that a radiological exam is performed, the higher the risk of deleterious effects to the irradiated fetus. Fortunately, Brown et al.[482] have shown that frequency of x-ray examination was significantly less during pregnancy than that of all women in the child bearing age. Furthermore, only approximately 21% of those women x-rayed during pregnancy were examined during the first trimester.

Unfortunately, many of the first trimester examinations are performed at a time when pregnancy is not known or even suspected. Examination of women known to be pregnant should be delayed as long as possible. Jacobsen[483] has published a document entitled, Instructions for X-Ray Examination of Pregnant Women and Women in the Reproductive Age Group as Applied at Cophenhagen County Hospital, Glostrup. When these instructions were first put into use, there was apparently considerable opposition to them by nonradiological hospital staff. However, as Jacobsen points out:

> "After a thorough discussion of the practical and theoretical problems at radiological conferences and at staff meetings, however, appreciation and readiness to cooperate were obtained from all departments. The work now runs smoothly."

The approach of the ICRP has been stated as follows: [317]

> "The commission has pointed out that the 10-day interval following the onset of menstruation is the time when it is most improbable that such women could be pregnant. Therefore, it is recommended that all radiological examinations of the lower abdomen and pelvis of women of reproductive capacity that are not of importance in connection with the immediate illness of the patient, be limited in time to this period when pregnancy is improbable. The examinations that it will be appropriate to delay until the onset of the next menstruation are the few that could without detriment be postponed until the conclusion of the pregnancy, or at least until its latter half."

These recommendations have been the subject of a number of recent papers.[484-488] Generally, these articles support the intent of the recommendations, but point out the numerous difficulties involved in actual clinical application.[484-496] Some of these administrative difficulties have reportedly been overcome[487-488] so that application of the ten-day rule can apparently run smoothly.

The irradiation of pregnant women may soon impose a more difficult and more frequent question to radiologists. The question is—when is therapeutic abortion indicated following a radiological examination during pregnancy? This situation has been alluded to in some publications and will no doubt become more important with our increasing liberalization of abortion laws.[489-493] Firm, numerical recommendations have not been generated for this situation and they should be. The diagnostic and the therapeutic radiologist are in need of guidance for this problem. In spite of an occasional report[494] of normal childbirth following high dose of radiation to the fetus, the overwhelming evidence suggests that pregnancy should be terminated following a fetal dose of 25 rads and perhaps lower, depending upon the time of irradiation and some nonmedical consideration.

8.2.4. Other Aspects of Procedural Control

Many of the previously mentioned administrative aspects of radiation control have been listed and discussed in stepwise fashion.[367,495] Administrative control procedures directed towards the improvement of radiographic quality are helpful in reducing the number of repeat examinations therefore reducing radiation exposures. Dawson[496] has described a method of compiling exposure tables to provide more precise exposure factors for a given patient physique. He employs on-line computer techniques for the specification of exposure factors and recommends this approach for other departments who have the computer capability. Computer applications are also currently under investigation to allow the computer to assist the radiologist in interpreting the film and to assist the administrator in the smooth operation of the department.[497] Future computer applications to image analysis, for example, promise to assist the radiologist in many routine film scans. Studies of image contrast enhancement[498,499] and darkroom practices[500] and time and motion studies[501] in departments of radiology are other examples of sophisticated investigations designed to increase technical and administrative efficiency in radiology and thereby contribute to effective radiation control.

One very good example of altered administrative procedure that has resulted in considerably reduced exposures is that dealing with mass miniature photofluorography. The excessive radiation exposure necessary for this type of examination has been known for a long time;[502,503] only recently have community public health agencies restricted the use of this technique.[504] Classically, the use of photofluorography has been primarily a mobile mass tuberculosis screening x-ray program. Tizes and Tizes[504] indicate that as of January 1970, 20 to 45 states replying to a questionnaire had discontinued their mobile photofluorographic program. The primary reason for discontinuance was that the yield of active cases of tuberculosis or other conditions was very small for the effort, expense, and excessive radiation required. No doubt, all routine chest examinations in some population groups, such as expectant mothers, will be discontinued.

8.2.5. Training and Education

One effective way to maintain proper radiation control is to increase the level of education and training for all radiological personnel. Higher qualifications of radiological personnel would not only result in less unnecessary exposure but also, no doubt, in increased diagnostic information.

At the National Conference on X-Ray Technician Training[505] it was pointed out that not only is there a shortage of radiological technicians but also the shortage is increasing as the demand for radiological services increases. This situation allows many unqualified personnel to gain employment as x-ray technologists, particularly in the private offices of nonradiologists. Legislation has been submitted to Congress to deal with this problem.[506] In effect the legislation requires that all operators of x-ray equipment be certified as competent from the radiation control standpoint. The legislation also allows for various classifications of x-ray operation that define limitations to the radiological procedures for which a technician is qualified. It is hoped that the stratification of technicians will encourage more to look upon x-ray technology as a career and thereby help to control the rapid turnover of technicians that many departments experience.

The need for closer attention to training and education of radiological personnel, including radiologists, was first alluded to by the National Advisory Committee on Radiation (NACOR) in their 1959 report.[507] Their 1966 report[508] states that, "It appears that the number of radiologists needed in the United States is almost twice as great as the number actually available." More recently the recommendations of NACOR have been expanded upon by the Committee on the Study of Academic Radiology of the National Academy of Sciences-National Research Council.[509] The recommendations relative to the training of radiologists made by this committee were enumerated as follows:

1. Increase the number of training positions in the university departments of radiology capable of providing well-rounded training.
2. Establish separate training programs in each of the major divisions of radiology.
3. Encourage the development of subspecialty training programs.
4. Establish educational and training (or retraining) programs in the allied sciences and professions, including radiation physics, radiobiology, and radiologic engineering.
5. Establish programs of research and development in radiologic education.

6. Increase the effectiveness of radiologists, as well as their numbers.

7. Establish grants for research facilities.

Some of the recommendations are currently being put into effect. The American College of Radiology is phasing out its recognition of general radiology residency programs and is encouraging the development of intensive training programs in radiological subspecialties, such as pediatric radiology, neuroradiology, and cardiovascular radiology. Other proposals have been submitted for a continuing certification of radiologists based not only on their medical ability but also on their knowledge of radiation control procedures.[510] These proposals suggest that new nonradiologist physicians be required to take licensure tests when they buy an x-ray machine, while currently practicing nonradiologist physicians would be permitted to continue to take x-rays as long as they used registered x-ray machines. Board certified radiologists would not be directly affected under this plan, but non-Board certified radiologists would be required to demonstrate competence in radiation control.

9. SUMMARY

This review has followed the development of our current radiation protection standards and radiation control practices in diagnostic Roentgenology from the time of Roentgen's discovery to the present. The development has spanned several eras of Roentgenological experience which are distinct, yet difficult to identify chronologically.

The first era covers the time of the Roentgen pioneers through approximately 1915. During this time, x-rays were a novelty in the hands of physicists and physicians. Many radiation injuries, including deaths, were reported and untold numbers of injuries were unreported. For the most part, the injuries were unreported. For the most part, the injuries were acute, easily observed, and obviously caused, and included most frequently skin erythema, desquamation, and epilation. The most severe injuries were malignant changes that often led to disfiguring surgical operations and death. During this time, only low kVp and low mA equipment was available which resulted in excessive exposure times and excessive differential patient dose relative to film exposure.

The second era covers the time period from approximately 1915 to the 1940's. It was during this time that radiation protection standards were first developed in an official manner and that radiation control practices were established at the local level and were reported. These standards and practices were instituted initially because of the early radiation injuries and the martyrdom that befell so many early Roentgen pioneers. With the institution of high capacity x-ray machines, reports of more subtle x-ray injuries appeared in the literature. These reports were primarily associated with the hematologic system and ranged from transient leukopenia to leukemia.

The radiation standards and practices of this second era were established on the assumption of a threshold-type dose-response relationship. It was generally accepted that if one worked within the recommended levels of exposure, no deleterious effects would ensue to either the patient or the operator. The emphasis of radiation control procedures was still focused on maintaining acceptable exposures to the individuals involved in each examination and at each facility. Radiation research directed to genetic damage resulted in linear nonthreshold type dose-response relationships. These findings produced suspicion that other effects were also nonthreshold which led to a reevaluation of the theoretical basis for radiation protection. Many radiation scientists became convinced that the basis for radiation control at that time was false. If, in fact, long-term latent effects followed a linear nonthreshold type relationship, then radiation exposures should be maintained at an even lower level than was at that time acceptable. The nonthreshold hypothesis gained acceptance and continues to the present time.

The final era of time covered by this review extends from the 1940's to the present day. During this time, acceptable radiation exposure levels have been reduced still further. The concern for radiation injury is no longer primarily directed to identifiable individuals, but rather to society at large, or at least population groups within society. Scientific reports relating identifiable human radiation injuries to diagnostic Roentgenological activities have been virtually absent for 20 years. They have been replaced, however, by large-scale epidemiological investigations that have attempted to show effects at the population level following diagnostic levels of radiation. The current radiation protection standards and radiation control practices are based on consideration of not only the earlier human injuries but also the current epidemiological evidence. There is no tolerance dose for the human radiation effects of concern today. Even the smallest radiation exposure is presumed to be accompanied by a finite probability of latent physiological effect. The value of this probability factor is unknown and can only be extrapolated from either high dose radiation studies or large-scale statistical analysis following low dose experiences.

Hopefully, it is clear that today's radiation control procedures and standards are very conservatively established. The radiation doses received by patients and personnel are extremely low and carry with them a negligible and undetectable probability of injury.

Today's programs for radiation control are designed primarily out of consideration for long-term population effects. The application of ionizing radiation by the radiologists should be made with caution, but also with a knowledgable understanding that radiation effects of an individual nature are highly improbable. However, since the magnitude of the effects of a population nature remain unknown, the application of x-rays in diagnosis should be conducted with prudence. Unquestionably, the benefits derived by society in the use of diagnostic x-rays outweigh the risk involved by several orders of magnitude. Still, it may be possible to maintain the present measure of benefit, while at the same time, reducing the measure of risk by conscientious application of accepted radiation control practices.

ARTICLES REVIEWED

Brown, M. L. et al., Population dose from X-rays, U. S. 1964, *U. S. Public Health Service*, #2001, 82, 1969.

Cantril, S. T., Biological bases for maximum permissible exposures, *Industr. Med. on the Plutoniun Project*, 20, McGraw Hill, New York, 1951.

Gitlin, J. N. and Lawrence, P. S., Population exposure to X-rays, U. S. 1964, Public Health Service Publication #1519, 1966.

Graham, S., Levin, M. L., Lihenfeld, A. M., Schuman, L. M., Gibson, R., Down, J. E., and Haempelman, W. Preconception, intrauterine and postnatal irradiation as related to leukemia, *Nat. Cancer Inst.*, Monograph #19, 347, 1966.

Hickey, P. M., The first decade of American Roentgenology, *Amer. J. Roentgen.*, 20, 150, 1928.

Leddy, E. T., Cilley, E. I. L., and Kirklin, B. R., The dangers of Roentgenoscopy and methods of protection against them. I. General review of the problem, *Amer. J. Roentgen.*, 32, 360, 1934.

Pfahler, G. E., Protection in radiology, *Amer. J. Roentgen.*, 9, 803, 1922.

Stewart, A. M., Radiogenic cancers in childhood, *Radiation Biology of the Fetal and Juvenile Mammal, Conf. 690501*, 681, 1969.

Stone, R. S., The concept of a maximum permissible exposure, *Radiology*, 58, 639, 1952.

Stone, R. S., Health protection activities of the plutonium project, *Proc. Amer. Phil. Soc.*, 90, 11, 1946.

Van Cleave, C. D., Radiation leukemogenesis, in *Late Somatic Effects of Ionizing Radiation*, TID 24310, 1968, 78.

REFERENCES

1. Grigg, E. R. N., *The Trail of the Invisible Light*, Charles C Thomas, Springfield, Ill., 1965.

2. Brecher, Ruth and Edward, *The Rays: A History of Radiology in the United States and Canada*, Williams & Wilkins Co., Baltimore, 1969.

3. Roentgen, W. C., On a new kind of rays, Sitzsber, *Phys.-Med. Ger-Wurzburg,* December 28, 1895.

4. Roentgen, W. C., On a new kind of rays - Second Communication, *Phys.-Med. Ger-Wurzburg,* 1896.

5. Kaye, G. W. C., *Roentgenology: Its Early History, Some Basic Physical Principles and the Protection Measures*, Paul B. Hoeber, Inc., Harper & Brothers, New York, 1928.

6. Glasser, Otto, *Wilhelm Conrad Roentgen and the Early History of the Roentgen Ray,* Charles C Thomas, Springfield Ill., 1934.

7. Glasser, Otto, *Dr. W. C. Roentgen*, 2nd ed., Charles C Thomas Springfield, Ill., 1958.

8. Glasser, O., Genealogy of roentgen rays, *Amer. J. Roentgen,* 30, 180, 1933.

9. Crane, A. W., The research trail of the X-ray, *Radiology*, 23, 131, 1934.

10. Smyth, H. D., From X-rays to nuclear fission, *Amer. Sci.,* 35, 485, 1947.

11. Morgan, Karl Z., Ionizing radiation: benefits versus risks, *Health Phys.,* 17, 539, 1969.

12. Weaver, Warren, What you should know about danger from X-rays, *U. S. News and World Report,* June 29, 1956.

13. Maisel, Albert Q., What is the truth about danger in X-rays?, *Reader's Digest,* February 1958.

14. Roentgen, W. C., On a new kind of rays, *Nature*, 53, 274, 1896.

15. Konig, F., The Roentgen pictures and the new installation for the physical department of the Institute of the Frankfort Physical Society, *User. Phys. Ver.,* Frankfort am, 1895/96.

16. Pupin, Michael, *From Immigrant to Inventor,* Charles Scribner Sons, New York, 1923, 306.

17. Hammeter, J. C., Photography of the human stomach by the Roentgen method, *Boston Med. J.,* 134, 609, 1896.

18. Hickey, P. M., The first decade of American Roentgenology, *Amer. J. Roengten.,* 20, 150, 1928.

19. Edison. T. A., Effect of x-rays upon the eye, (note) *Nature,* 53, 421, 1896.

20. Walsh, David, Deep tissue traumatism from Roentgen ray exposure, *Brit. Med. J.,* 2, 272, 1897.

21. Colwell, H. A. and Russ, S., *X-ray and Radium Injuries: Prevention and Treatment,* London Oxford University Press, 1934.

22. Stone, R. S., Fifty years of radiology: From Roentgen to the era of atomic power, *Western J. Surg.*, 54, 153, 1946.

23. Rolleston, H., Harmful effects of irradiation: A critical review, *Quart. J. Med.,* 4, 101, 1930.

24. Grubbe, E. H., *X-Ray Treatment, Its origin, Birth and Early History,* Bruce Publishing Co., St. Paul, Minn., 1949.

25. Kathren, L., Early X-ray protection in the United States, *Health Phys.*, 8, 503, 1962.

26. *J. Roentgen Soc.*, 12, 67, 1916.

27. Stone, R. S., The concept of a maximum permissible exposure, *Radiology*, 58, 639, 1952.

28. Rollins, W., Notes on X-ray light: Vacuum tube burns, *Boston Med. Surg. J.*, 146, 39, 1902.

29. Rollins, W., *Notes on X-light*, privately published, Boston, 1904.

30. Winkler, K. G., Influence of rectangular collimation and intraoral shielding on radiation dose in dental radiography, *J. Amer. Dent. Ass.*, 77, 95, 1968.

31. Laurence, W. S., Discussion, *Trans. Amer. Roentgen Ray Soc.*, 117, 1907.

32. Caldwell, E. W., Discussion, *Trans. Amer. Roentgen Ray Soc.*, 111, 1907.

33. Russ, S., The injurious effects caused by X-rays, *J. Roentgen Soc.*, 12, 38, 1916.

34. Russ, S., Hard and soft X-rays, *Arch. Roentgen Ray,* (London), 19, 323, 1914.

35. Kaye, G. W. C., *X-rays: An Introduction to the Study of Roentgen Rays,* Longmans-Green & Co., London, 1918, Appendix V.

36. Resolution of the Roentgen Society, *J. Roentgen Soc.*, 12, 68, 1916.

37. British X-ray and Radium Protection Committee, X-ray and radium protection, *J. Roentgen Soc.*, 17, 100, 1921.

38. Rolleston, H. (Chairman), Preliminary report of the x-ray and radium protection committee, *Arch. Radiol. Electrother.*, 26, 5, 1921.

39. Cantril, S. T., Biological basis for maximum permissible exposures, *Industr. Med. on the Plutonium Project,* 20, McGraw Hill, New York, 1951.

40. Desjardins, A. U., Protection against radiation, *Radiology*, 1, 221, 1923.

41. Mutscheller, A., Physical standards of protection against Roentgen ray dangers, *Amer. J. Roentgen.*, 13, 65, 1925.

42. Mutscheller, A., Safety standards of protection against X-ray dangers, *Radiology*, 10, 468, 1928.

43. Mutscheller, A., More on X-ray protection standards, *Radiology,* 22, 739, 1934.

44. Pfahler, G. E., Protection in radiology, *Amer. J. Roentgen.*, 9, 803, 1922.

45. Carman, R. D. and Miller, A., Occupational hazards to the radiologist with special reference to change in the blood, *Radiology*, 3, 408, 1924.

46. Risse, A., Some blood changes during Roentgen and radium action, Internat. Med. and Surg. Survey, (X) *Roentgenology*, Vii 127, 1924.

47. Lacassagne, A. and Lavedan, J., Histologic modifications of the blood after experimental radiations, *Internat. Med. and Surg. Survey,* (X) *Roentgenology,* Vii, 269, 1924.

48. Klein, J., The rapidity of sedimentation of erythrocytes before and after Roentgen irradiation, *Internat. Med. and Surg. Survey,* (X) *Roentgenology*, Vii, 128, 1924.

49. Barclay, A. S. and Cox, S., The radiation risks of the Roentgenologist, *Amer. J. Roentgen.,* 19, 551, 1928.

50. Failla, G., Radiation protection, *Radiology*, 19, 12, 1932.

51 Lazarus-Barlow, W. S., Some biological effects of small quantities of radium, *Arch. Radiol. Electrother.*, XXIV, 1, 1919,

52. Kaye, G. W. C., Bell, G. E., and Binks, W., Protection of radium workers from gamma radiation, *Brit. J. Radiol,* 8, 6, 1935.

53. The dangers in working with radium and X-rays. How the problem is being met, *Radium*, 17, 4, 1921.

54. Sievert, R. M., Einige Untersuchungen ueber Vorrichtungen zum Schutz gegen Roentgenstrahlen, *Acta Radiol.*, 4, 61, 1925.

55. Kaye, G. W. C., Some fundamental aspects of Roentgen rays and the protection of the Roentgen ray worker, *Amer. J. Roentgen.*, 18, 401, 1927.

56. Leddy, E. T., Cilley, E. I. L., and Kirklin, B. R., The dangers of Roentgenoscopy and methods of protection against them. I. General review of the problem, *Amer. J. Roentgen.*, 32, 360, 1934.

57. Cilley, E. I. L., Leddy, E. T., and Kirklin, B. R., The dangers of Roentgenoscopy and methods of protection against them. II. Some considerations of the size of the beam used in Roentgenoscopic examination, *Amer. J. Roentgen.*, 32, 805, 1934.

58. Cilley, E. I. L., Leddy, E. T., and Kirklin, B. R., The dangers of Roentgenoscopy and methods of protection against them. III. The protective power of the barium-filled stomach, *Amer. J. Roentgen.*, 33, 88, 1935.

59. Cilley, E. I. L., Kirklin, B. R., and Leddy, E. T., The dangers of Roentgenoscopy methods of protection against them. IV. A detailed consideration of the doses received by the fingers of the examiner, *Amer. J. Roentgen.*, 33, 390, 1935.

60. Cilley, E. I. L., Leddy, E. T., and Kirklin, B. R., The dangers of Roentgenoscopy and methods of protection against them. V. Some considerations of the "dose" received during examination of the colon, *Amer. J. Roentgen.*, 33, 787, 1935.

61. Turnbull, A. and Leddy, E. T., A method for determining the limits of safety in Roentgenography, *Amer. J. Roentgen.*, 34, 258, 1935.

62. Cumberbatch, E. P., Fatal leucopenia following X-ray treatment, *Arch. Roentgen. Ray*, 18, 187, 1913.

63. Collear, T. E., Kodak and radiography, *Med. Radiogr. Photogr.*, 46, 78, 1970.

64. International Congress of Radiology Recommendations, X-ray and radium protection, *Radiology*, 12, 519, 1929.

65. National Bureau of Standards Circular #374, 1929.

66. Medical X-ray Protection up to Two-Million Volts, *National Bureau of Standards Handbook* 41, 1949.

67. International recommendations for X-ray and radium protection, *Brit. J. Radiol.*, 5, 82, 1932.

68. International recommendations for X-ray and radium protection, *Radiology*, 23, 682, 1934.

69. International recommendations for X-ray and radium protection, *Radiology*, 30, 511, 1938.

70. X-Ray protection, *National Bureau Standards Handbook* #15, 1931.

71. Radium Protection, *National Bureau Standards Handbook* #18, 1934.

72. X-Ray Protection, *National Bureau Standards Handbook* #20, 1936.

73. Kaye, G. W. C., *Roentgenology,* Paul B. Holber, New York, 1928.

74. Wintz, L. and Rump, A., *League of Nations Publication CH-1054*, 1931.

75. Taylor, L. X-Ray protection, *J. Amer. Med. Ass.*, 116, 135, 1941.

76. Paterson, R., Effects of radiation on workers, *Brit. J. Radiol.*, 16, 2, 1943.

77. Parker, H. M., Health Physics, Instrumentation and Radiation Protection, *Advances Biol. Med. Phys.* 1, 223, 1948.

78. Kaye, G. W. C., The emission and transmission of Roentgen rays, *Phil. Trans. Roy. Soc.*, A. 209, 123, 1908.

79. Editorial, *Amer. J. Roentgen.*, 1, 90, 1913.

80. Kaye, G. W. C., *X-Rays: An Introduction to the Study of Roentgen Rays*, Longmans-Green & Co., London, 1918.

81. Cole, L. G., A preliminary report on the diagnostic and therapeutic application of the Coolidge tube, *Amer. J. Roentgen.*, 1, 125, 1914.

82. Jerman, E. C., Roentgen-ray apparatus, *The Science of Radiology*, Glasser, O., Ed., Charles C Thomas, Springfield, Ill., 1933, 64.

83. Shearer, J. S., Possible dangers in connection with the use of the X-rays and how to avoid them, *Amer. J. Roetgen.*, 10, 240, 1923.

84. Wilsey, R. B., Some practical results with a Potter-Bucky diaphragm, *Amer. J. Roentgen.*, 9, 441, 1922.

85. Potter, H. E., History of diaphragming Roentgen rays by use of the Bucky Principle, *Amer. J. Roentgen.*, 25, 396, 1931.

86. Muller, Hermann J., The effect of Roentgen rays upon the hereditary material, in *The Science of Radiology*, Glasser, O., Ed., Charles C Thomas, Springfield, Ill., 1933, 305.

87. Warthin, A. S., An experimental study of the effects of Roentgen rays upon the blood forming organs, with special reference to the treatment of leukemia, *Int. Clin.*, 15, 243, 1906.

88. Desjardins, A. U. and Marquis, W. J., Blood count and blood pressure in high voltage therapy, *Radiology*, 2, 252, 1924.

89. Cori, C. F., Biological reaction of X-rays: The influence of X-ray treatment on the complement content of the blood of cancer patients, *Amer. J. Roentgen.*, 10, 830, 1923.

90. Hirsch, E. F. and Peterson, A. J., The blood with deep Roentgen-ray therapy: Hydrogen ion concentration, alkali reserve, sugar and non-protein nitrogen, *J. A. M. A.*, 80, 1505, 1923.

91. Taylor, Herbert D., Witherbee, William, D., and Murphy, James B., Studies on X-ray effects. I. Destructive action on blood cells, *J. Exp. Med.*, 29, 53, 1919.

92. Thomas, M. M., Taylor, H. D., and Witherbee, W. D., Studies on X-ray effects. II. Stimulative action of the lymphocytes. *J. Exp. Med.*, 29, 75, 1919.

93. Nakahara, W., Studies on X-ray effects. III. Changes in the lymphoid organs after small doses of X-rays, *J. Exp. Med.*, 29, 83, 1919.

94. Russ, S., Chambers, H., and Scott, G. M., Further observations of the effects of X-rays upon lymphocytes, *Arch. Radiol.*, 25, 197, 1921.

95. Mottram, J. C., Histological changes in the bone marrow of rats exposed to the gamma radiations from radium, *Arch. Radiol.*, 25, 197, 1921.

96. Levin, I., Action of radium and the X-rays on the blood and blood forming organs, *Amer. J. Roentgen.*, 9, 112, 1922.

97. Falconer, E. H., Morris, L. M., and Ruggles, H. E., The effect of X-rays on bone marrow, *Amer. J. Roentgen.*, 11, 342, 1924.

98. Krebs, C., Rask-Nielsen, H. D., and Wagner, A., The origin of lymphosarcomatosis and its relation to other forms of leukosis in white mice, *Acta Radiol.*, Suppl. 10, 1, 1930.

99. Furth, J., Transmission of myeloid leukemia in mice, *Proc. Soc. Exp. Biol. Med.*, 31, 923, 1934.

100. Hueper, W. C., Leukemoid and leukemic conditions in white mice with spontaneous mammary carcinoma, *Folia Haemat.*, 52, 167, 1934.

101. Furth, J., Rathbone, R. R., and Seibold, H. R., Relation of X-rays to lymphomatosis, *Proc. Soc. Exp. Biol. Med.*, 30, 754, 1933.

102. Furth J., Seibold, H. R., and Rathbone, R. R., Experimental studies on lymphomatosis of mice, *Amer. J. Cancer*, 19, 521, 1933.

103. Leitsch, A., The immediate effects of X-rays on the blood lymphocytes, *Arch. Radiol.*, 26, 122, 1921-1922.

104. Russ, S., The immediate effects of X-rays on the blood lymphocytes, *Arch. Radiol.*, 26, 146, 1921-1922.

105. Portis, M. D., Blood changes in workers with the Roentgen ray and apparatus for protection, *J. A. M. A.*, 65, 20, 1915.

106. Mottram, J. C., and Clarke, J. R., The leukocytic blood content of those handling radium for therapeutic purposes, *Arch. Radiol.*, 24, 345, 1919-1920.

107. Mottram, J. C., The red cell blood content of those handling radium for therapeutic purposes, *Arch. Radiol.*, 25, 194, 1920.

108. Mottram, J. C., The use of blood counts to indicate the efficiency of X-ray and radium protection, *Brit. Med. J.*, 269, 1921.

109. Pfahler, G. E., The effects of the X-rays and radium on the blood and general health of radiologists, *Amer. J. Roentgen.*, 9, 647, 1922.

110. Mottram, J. C., The effect of increased protection from radiation upon the blood conditions of radium workers, *Arch. Radiol.*, 26, 368, 1921.

111. Amundsen, P., Blood anomalies in radiologists and in persons employed in radiological service, *Acta Radiol.*, 3, 1, 1924.

112. Goodfellow, D. R., Leucocytic variations in radium workers *Brit. J. Radiol.*, 8, 669, 1935.

113. Goodfellow, D. R., Leucocytic variations in radium workers (Part II), *Brit. J. Radiol.*, 8, 752, 1935.

114. Larkins, F. E., A case of acute aplastic anemia, *Arch. Radiol.*, 25, 380, 1921.

115. Loutit, J. R., Radiation exposure of staff in diagnostic procedures, *Brit. J. Radiol.*, 28, 647, 1955.

116. Stone, R. S., Health protection activities of the plutonium project, *Proc. Amer. Phil. Soc.*, 90, 11, 1946.

117. Jacobson, L. O. and Marks, E. K., The hematologic effects of ionizing radiations in the tolerance range, *Radiology*, 49, 286, 1947.

118. Moshman, J., Hematological effects of chronic low-level irradiation, *J. Appl. Physiol.*, 4, 145, 1951.

119. Hunter, F. T., Merrill, O. E., Trump, J. G., and Robbins, L. L., Protection of personnel engaged in Roentgenology and radiology, *New Eng. J. Med.*, 241, 79, 1949.

120. Hunter, F. T. and Robbins, L. L., Protection of personnel engaged in Roentgenology and radiology: Final report, *New Eng. J. Med.*, 244, 9, 1951.

121. National Committee on Radiation Protection and Measurements, Statement on blood counts, *Radiology*, 63, 428, 1954.

122. Nuttall, John R., The effects of occupational exposure to X-rays and radioactive substances, *Clin. J.*, 72, 181, 1943.

123. Mayneord, W. V., Some problems of radiation protection, *Brit. J. Radiol.*, 24, 525, 1951.

124. Paterson, R., Effects of radiation on workers, *Brit. J. Radiol.*, 16, 2, 1943.

125. Wald, N., Thoma, G. E., and Broun, G., Hematologic manifestations of radiation exposure in man, *Progr. Hemat.*, 3, 1, 1962.

126. Kelley, S. and Brown, C. D., Chromosome aberrations as a biological dosimeter, *Amer. J. Public Health*, 54, 1419, 1964.

127. Heddle, John A., Radiation induced chromosome aberrations in man, a possible biological dosimeter, *Fed. Proc.*, 28, 1790, 1969.

128. Nowell, Peter C., Biological significance of induced human chromosome aberrations, *Fed. Proc.*, 28, 1797, 1969.

129. Scharpe, H., Pitfalls in the use of chromosome aberration analysis for biological radiation dosimetry, *Brit. J. Radiol.*, 42, 943, 1969.

130. El-Alfi, O. S., Ragab, A. S., and Eassu, E. M., Detection of radiation damage in exposed personnel by chromosome study, *Brit. J. Radiol.*, 40, 760, 1967.

131. Wald, N., Feagin, F., and Ranshaw, R., Automation of human cytogenetic study methodology, *Pro. 6th Int. Conf. Med. Elec. Biol. Engr.*, 152, 1965.

132. Butler, J. W. and Butler, M. K., Computer analysis of photographic images, *Nucleonics*, 25, 44, 1967.

133. Lewis, E. B., Leukemia and ionizing radiation, *Science*, 125, 965. 1957.

134. Van Cleave, C. D., Radiation leukemogenesis, in *Late Somatic Effects of Ionizing Radiation*, TID 24310, 1968, 78.

135. Dunlap, C. E., Effects of radiation on normal cells. III. Effects of radiation on the blood and the hemopoietic tissues, including the spleen, the thymus and the lymph nodes, *Arch. Path.*, 34, 562, 1942.

136. Henshaw, P. S. and Hawkins, J. W., Incidence of leukemia in physicians, *J. Nat. Cancer Inst.*, 4, 339, 1944.

137. Ulrich, H., The incidence of leukemia in radiologists, *New Eng. J. Med.*, 234, 45, 1946.

138. Dublin, L. I. and Spiegelman, M., Mortality of medical specialists, 1938-1942, *J.A.M.A.*, 137, 1519, 1948.

139. March, H. C., Leukemia in radiologists, correspondence, *J.A.M.A.*, 135, 179, 1947.

140. March, H. C., Leukemia in radiologists, *Radiology*, 43, 275, 1944.

141. March, H. C., Leukemia in radiologists in a 20-year period, *Amer. J. Med. Sci.*, 220, 282, 1950.

142. Warren, S., Longevity and causes of death from irradiation in physicians, *J.A.M.A.*, 162, 464, 1956.

143. Lewis, E. B., Leukemia, multiple myeloma, and aplastic anemia in American radiologists, *Science*, 142, 1492, 1963.

144. Warren, S. and Lombard, O. M., New data on the effects of ionizing radiation on radiologists, *Arch. Environ. Health*, 13, 415, 1966.

145. Folley, J. H., Borges, W., and Yamazaki, T., Incidence of leukemia in survivors of atomic bomb in Hiroshima and Nagasaki, Japan, *Amer. J. Med.*, 13, 311, 1952.

146. Lange, R. D., Moloney, W. C., and Yamazaki, T., Leukemia in atomic bomb survivors. I. General observations, *Blood*, 9, 574, 1954.

147. Moloney, W. C. and Lange, R. D., Leukemia in atomic bomb survivors. II. Observations on early phases of leukemia, *Blood*, 9, 663, 1954.

148. Moloney, W. C. and Kastenbaum, M. A., Leukemogenic effects of ionizing radiation on atomic bomb survivors in Hiroshima City, *Science*, 121, 308, 1955.

149. Moloney, W. C., Leukemia in survivors of atomic bombing, *New Eng. J. Med.*, 253, 88, 1955.

150. Heyssel, R. M., Brill, A. B., Woodbury, L. A., Nishimura, E. T., Ghose, T., Hoshino, T., and Yamasaki, M., Leukemia in Hiroshima atomic bomb survivors, *Blood*, 15, 313, 1960.

151. Brill, A. B., Tomonaga, M., and Heyssel, R. M., Leukemia in man following exposure to ionizing radiation: Summary of findings in Hiroshima and Nagasaki and comparison with other human experience, *Ann. Intern. Med.*, 56, 590, 1962.

152. Anderson, R. E., Yamamoto, T., Yamada, A., and Will, D. W., Autopsy study of leukemia in Hiroshima, *Arch. Path.*, 78, 618, 1964.

153. Ishimara, T., Hoshino, T., Ichimara, M., Okado, H., Tomigasu, T., Tsuchemoto, T., and Yamamoto, T., Leukemia in atomic bomb survivors, Hiroshima and Nagasaki. 1 October, 1950 - 30 September, 1966, *Radiat. Res.*, 45, 216, 1971.

154. Bizzozero, O. J., Johnson, K. G., and Ciocco, A., Radiation related leukemia in Hiroshima and Nagasaki, 1946-1964. *New Eng. J. Med.*, 274, 1095, 1966.

155. Report of the United Nations Scientific Committee on the Effects of Atomic Radiation, 19th Session, Suppl. #14, New York, 1964.

156. Court-Brown, W. M. and Doll, R., Leukemia and aplastic anemia in patients irradiated for ankylosing spondylitis, *Medical Research Council Special Report 295*, HMSO, London, 1957.

157. Hempelmann, L. H., Pifer, J. W., Burke, G. J., Terry, R., and Ames, W. R., Neoplasm in persons treated with X-rays in infancy for thymic enlargement. A report of the third followup study, *J. Nat. Cancer Inst.*, 38, 317, 1967.

158. Pochin, E. E., Leukemia following radioiodine treatment for thyrotoxicosis, *Brit. Med. J.*, 2, 1545, 1960.

159. Faber, M., Radiation induced leukemia in Denmark, in *Advances in Radiobiology*, de Heuesy, G. C., Forssberg, A. G., and Abbatt, J. D., Eds., Charles C. Thomas, Springfield, Ill., 1957, 397.

160. Simon, N., Brucer, M., and Hayes, R., Radiation and leukemia in carcinoma of the cervix, *Radiology*, 74, 905, 1960.

161. Gunz, F. W. and Atkinson, H. R., Medical radiations and leukemia. A retrospective survey, *Brit. Med. J.*, 1, 389, 1964.

162. Cronkite, E. P., Moloney, W., and Bond, V. P., Radiation leukemogenesis. An analysis of the problem. *Amer. J. Med.*, 28, 673, 1960.

163. Stewart, A., Pennybacker, W., and Barber, R., Adult leukaemias and diagnostic x-rays, *Brit. Med. J.*, 2, 882, 1962.

164. Stewart, A., Webb, J., and Hewitt, D., Survey of childhood malignancies, *Brit. Med. J.*, 1, 1495, 1958.

165. Neumann, G., Roentgen diagnosis and incidence of leukemia, *Deutsch Med. Wschr.*, 87, 90, 1962.

166. Polhemus, D. W. and Koch, R., Leukemia and medical radiation, *Pediatric*, 23, 453, 1959.

167. Birch, A. M. and Baker, D. H., Effect of repeated fluoroscopic examinations on 1480 children with long-term followup study, *New Eng. J. Med.*, 262, 1004, 1967.

168. Murray, R., Heckel, P., and Hempelmann, L. H., Leukemia in children exposed to ionizing radiation, *New Eng. J. Med.* 261, 585, 1959.

169. Graham, S., Levin, M. L., Lihenfeld, A. M., Schuman, L. M., Gibson, R., Down, J. E., and Hempelmann, L. H., Preconception, intrauterine and postnatal irradiation as related to leukemia, *Nat. Cancer Inst.* Monograph #19, 347, 1966.

170. Russell, L. B. and Russell, W. L., An analysis of the changing radiation response of the developing mouse embryo, *J. Cell. Physiol.*, 43, supp. 1, 103, 1954.

171. Yamazaka, J. N., A review of the literature on the radiation dosage required to cause manifest central nervous system disturbances from *in utero* and postnatal exposure, *Pediatrics*, 37, 877, 1966.

172. Rugh, R., Chairman's remarks, *Radiation biology of the fetal and juvenile mammal*, Conf. 690501, 381, 1969.

173. Van Cleave, C. D., Prenatal irradiation and late life vigor, in Late somatic effects of ionizing radiation, TID-24310, 1968, 176.

174. Sternglass, E. J., Cancer: Relation to prenatal radiation to development of the disease in childhood, *Science*, 140, 1102, 1963.

175. Sternglass, E. J., Infant mortality and nuclear tests, *Bull. At. Sci.*, 25, 18, April 1969.

176. Sternglass, E. J., A reply, *Bull. At. Sci.*, 25, 29, October 1969.

177. Sternglass, E. J., A reply, *Bull. At. Sci.* 25, 29, December 1969.

178. Sternglass, E. J., Evidence for low level radiation effects on the human embryo and fetus, *Radiation Biology of the Fetal and Juvenile Mammal*, Conf. 690501, 693, 1969.

179. Morgan, K. Z., Testimony, *Hearings before the Committee on Commerce*, U. S. Senate, Serial #90-49, 54, 1967.

180. Morgan, K. Z., Testimony, *Hearings before the Committee on Interstate and Foreign Commerce, House of Representatives*, Serial No. #90-11, 370, 1967.

181. Gofman, J. W. and Tamplin, A. R., Radiation, cancer and environmental health, *Hosp. Pract.*, 91, October 1970.

182. Tamplin, A. R., Fetal and infant mortality and the environment, *Bull. At. Sci.*, 23, December 1969.

183. Heller, M., Infant and fetal mortality caused by SR-90, *AAPM Quarterly*, 4, 24, 1970.

184. Bond, V. P., Radiation standards, particularly as related to nuclear power plants, *Health Phys. Soc. Newslett.*, Feburary 9, 1971.

185. Stewart, A., Webb, J., Giles, D., and Hewitt, D., Malignant disease in childhood and diagnostic irradiation *in utero*, *Lancet*, 2, 447, 1956.

186. Stewart, A. and Hewitt, D., Leukemia incidence in children in relation to radiation exposure in early life, *Curr. Top. Radiat. Res.*, 1, 221, 1965.

187. Stewart, A. and Kneale, G. W., Changes in the cancer risk associated with obstetric radiography, *Lancet*, 1, 104, 1968.

188. Stewart, A. M., Radiogenic cancers in childhood, *Radiation Biology of the Fetal and Juvenile Mammal, Conf. 690501*, 681, 1969.

189. Stewart, A. and Kneale, G. W., Radiation dose effects in relation to obstetric X-rays and childhood cancers, *Lancet*, 1185, June, 1970.

190. Ford, D., Paterson, J. C. S., and Treuting, W. L., Fetal exposure to diagnostic X-rays and leukemia and other malignant diseases in childhood, *J. Nat. Cancer Inst.*, 22, 1093, 1959.

191. Court-Brown, W. M., Doll, R., and Hill, A. B., The incidence of leukemia after exposure to diagnostic radiation *in utero*, *Brit. Med. J.*, 2, 1539, 1960.

192. MacMahon, B., Prenatal X-rays exposure and childhood cancer, *J. Nat. Cancer Inst.*, 28, 1173, 1962.

193. MacMahon, B., and Hutchison, G. B., Prenatal X-ray and childhood cancer, A review, *Union Int. Contra Cancrum Acta*, 20, 1172, 1964.

194. Kaplan, H. S., An evaluation of the somatic and genetic hazards of the medical uses of radiation, *Amer. J. Roentgen.*, 80, 696, 1958.

195. Kjldsberg, H., Relationship between leukemia and X-rays, *Tidsskr. nor. Laegeforen.*, 77, 1052, 1957.

196. Lewis, T. L. T., Leukemia in childhood after antenatal exposure to X-rays, *Brit. Med. J.*, 2,1551, 1960.

197. Murray, R., Heckel, P., and Hempelmann, L. H., Leukemia in children exposed to ionizing radiation, *New Eng. J. Med.*, 261, 585, 1959.

198. Wells, J. and Steer, C. M., Relationship of leukemia in children to abdominal irradiation, *Amer. J. Obstet. Gynec.*, 81, 1059, 1961.

199. Gunz, F. W., Bothwick, R. A., and Rolleston, G. L., Acute leukemia in an infant following excessive intrauterine irradiation, *Lancet*, 2, 19, 1958.

200. Jablon, Seymour and Kato, Hiroo, Childhood cancer in relation to prenatal exposure to atomic-bomb radiation, *Lancet*, 2, 1000, 1970.

201. Ager, E. A., Schuman, L. M., Wallace, H. M., Rosenfield, A. B., and Gullen, W. H., An epidemiological study of childhood leukemia, *J. Chronic Dis.*, 18, 113, 1965.

202. Buck, C., Population size required for investigating threshold dose in radiation induced leukemia, *Science*, 129, 1357, 1959.

203. International Commission on Radiological Protection, The evaluation risks from radiation, *Health Phys.*, 12, 239, 1966.

204. Auxier, J. A., Cheka, J. S., Haywood, F. F., Jones, T. D., and Thorngate, J. H., Free-field radiation-dose distributions from the Hiroshima and Nagasaki bombings, *Health Phys.*, 12, 425, 1966.

205. Marinelli, L. D., Estimates of the radiation-induced leukemic risk in man, *Argonne National Laboratory, Radiological Physics Division Annual Report*, July, 1969-June, 1970.

206. Wise, M. E., The latent period and its variation in human leukemia induced by X-rays, *Health Phys.*, 4, 250, 1961.

207. Gunz, F. W. and Atkinson, H. R., Leukemia following irradiation, *Radiology*, 84, 1067, 1964.

208. Sagan, L. A., Medical uses of radiation, *J.A.M.A.*, 215, 1977, 1971.

209. Socolow, E. L., Hashizame, A., and Neruski, S., Thyroid carcinoma in man after exposure to ionizing radiation. A summary of findings in Hiroshima and Nagasaki, *New Eng. J. Med.*, 268, 406, 1963.

210. Zeldis, J. L., Jablon, S., and Ishida, M., Current status of ABCC-NIH studies of carcinogenesis in Hiroshima and Nagasaki, *Ann. N. Y. Acad. Sci.* 114, 225, 1964.

211. Wanebo, C. K., Johnson, K. G., Sato, K. and Thorslund, T. W., Breast cancer after exposure to atomic bombings of Hiroshima and Nagasaki, *New Eng. J. Med.* 279, 667, 1968. 212.

212. Wanebo, C. K., Johnson, K. G., Sato, K., and Tharslund, T. W., Lung cancer following atomic radiation, *Amer. Rev. Resp. Dis.*, 98, 778, 1968.

213. Miller, Robert W., Delayed radiation effects in atomic-bomb survivors, *Science*, 166, 569, 1969.

214. Saenger, E. L., Silverman, F. N., and Sterling, T. D., Neoplasia following therapeutic irradiation for benign conditions in childhood, *Radiology*, 74, 889, 1960.

215. Tables XIV, SVII and SVIII in Report of the United Nations Scientific Committee on the Effects of Atomic Radiation, 19th Session, Suppl. #14, 1964.

216. Takahashi, S., Kitabatoki, T., and Wakabayashi, M., A statistical study on human cancer induced by medical exposures, *Nippon Acta Radiol.* 23, 1510, 1964.

217. Brinkley, Diana and Haybittle, J. L., The late effects of artificial menopause by X-radiation, *Brit. J. Radiol.*, 42, 519, 1969.

218. Raventos, A. and Winship, T., The latent interval for thyroid cancer following irradiation, *Radiology*, 83, 501, 1964.

219. Sagerman, R. H., Cassady, J. R., Tretter, P., and Ellsworth, R. M., Radiation induced neoplasia following external beam therapy for children with retinoblastoma, *Amer. J. Roentgen.*, 105, 529, 1969.

220. Hemplemann, L. H., Risk of thyroid neoplasm after irradiation in childhood, *Science*, 160, 159, 1968.

221. Lindop, P. J., and Sacher, G. A., *Radiation and Aging*, Taylor and Francis, Ltd., London, 1966.

222. Curtis, H. J., Radiation induced aging, in *Medical Physics*, Vol. 3, Glasser, O., Ed., Year Book Publisher, Inc., Chicago, 1960, 492.

223. Storer, J. B., Radiation resistance with age in normal and irradiated populations of mice, *Radiat. Res.*, 25, 435, 1965.

224. Grahn, D. and Sacher, G. A., Fractionation and protraction factors and the late effects of radiation in small animals, in *Dose Rate in Mammalian Radiation Biology*, Brown, D., Gragle, R., and Noonan, T., Eds., Conf. 680410, 1968, Chap. 2.

225. Gowen, J. W. and Stadler, J., Life spans of different strains of mice as affected by acute irradiation with 100 kVp X-rays, *J. Exp. Zool.*, 132, 133, 1956.

226. Grahn, D. and Sacher, G. A., Chronic radiation mortality in mice after single whole-body exposure to 250-, 135-, and 80-kVp X-rays, *Radiat. Res.*, 8, 187, 1958.

227. Grahn, D., Genetic control of physiological processes: The genetics of radiation toxicity in animals, in *Radioisotopes in the Biosphere*, Caldecott, R. S. and Snyder, L. S., Eds., Minnesota, 1960, 181.

228. Upton, A. C., Kimball, A. W., Furth, J., Christenberry, K. W., and Benedict, W. H., Some delayed effects of atom-bomb radiations in mice, *Cancer Res.*, 20, 1, 1960.

229. Lindop, P. J. and Rotblat, J., Long-term effects of a single whole-body exposure of mice to ionizing radiations. I. Life-shortening, *Proc. Roy. Soc. Med.*, 154, 332, 1961.

230. Storer, J. B., Evaluation of radiation response as an index of aging in mice, *Radiat. Res.*, 17, 878, 1962.

231. Spalding, J. F., Johnson, O. S., and McWilliams, P. C., Dose rate-total dose effect from single short-duration gamma-ray exposures on survival time in mice, *Radiat. Res.*, 32, 21, 1967.

232. Upton, A. C., Randolph, M. L., and Conklin, J. W., Late effects of fast neutrons and gamma rays in mice as influenced by the dose rate of irradiation: Life shortening, *Radiat. Res.*, 32, 493, 1967.

233. Dublin, L. I. and Spiegelman, M., The longevity and mortality of American physicians, 1938-1942. A preliminary report, *J.A.M.A.*, 134, 1211, 1947.

234. Seltzer, R. and Sartwell, P. E., Ionizing radiation and Longevity in physicians, *J.A.M.A.*, 166, 585, 1958.

235. Warren, S., The basis for the limit on whole-body exposure – Experience of radiologists, *Health Phys.*, 12, 737, 1966.

236. Seltser, R. and Sartwell, P. E., The influence of occupational exposure to radiation on the mortality of American radiologists and other medical specialists, *Amer. J. Epidem.*, 81, 2, 1965.

237. Beebe, G. W., Ishida, M., and Jablon, S., Studies of the mortality of A-bomb survivors. I. Plan of study and mortality in medical subsample (selection 1) 1950-1958, *Radiat. Res.*, 16, 253, 1962.

238. Jablon, S., Ishida, M., and Beebe, G. W., Studies of the mortality of A-bomb survivors. 2. Mortality in selections I and II, 1950-1959, *Radiat. Res.*, 21, 423, 1964.

239. Jablon, S., Ishida, M., and Yamsaki, M., Studies of the mortality of A-bom survivors. 3. Description of the sample and mortality, 1950-1960, *Radiat. Res.*, 25, 25, 1965.

240. Court-Brown, W. M. and Doll, R., Expectation of life and mortality from cancer among British radiologists, *Brit. Med. J.*, 2, 181, 1958.

241. Glass, R. L., Mortality of New England Dentists 1921-1960, U. S. Department of Health, Education and Welfare, *Public Health Service Publication* #999-RH-18, 70, 1966.

242. Miller, R. W. and Jablon, S., A search for late radiation effects among men who served as X-ray technologists in the U. S. Army during World War II, *Radiology*, 96, 269, 1970.

243. Griem, M. L., Meier, P., and Dobben, Glen, Analysis of the morbidity and mortality of children irradiated in fetal life, *Radiology*, 88, 347, 1967.

244. Jones, H. B., Factors in longetivity, *Kaiser Foundation Med. Bull.*, 4, 329, 1956.

245. Failla, G. and McClement, P., The shortening of life by chronic whole body irradiation, *Amer. J. Roentgen.*, 78, 946, 1957.

246. Lorenz, E., Biological effects of external x and gamma radiation, in *National Nuclear Energy Series N-22B.*, Zinkle, R. E., Ed., McGraw-Hill, New York, 1954, 24.

247. Langham, W., The radiobiological factors in manned space flight, Nat. Acad. Sci. Nat. Res. Counc., 1967.

248. Abelson, P. H. and Kruger, P. G., Cyclotron induced radiation cataracts, *Science*, 110, 655, 1949.

249. Britten, M. J. A., Halnan, Keith, E., and Meredith, W. J., Radiation cataract – new evidence on radiation dosage to the lens, *Brit. J. Radiol.*, 39, 612, 1966.

250. Hickey, Preston M., A report analyzing the results of the questionnaire sent out to radiologists under the direction of the sex committee of the National Research Council, *Amer. J. Roentgen.*, 18, 458, 1927.

251. Macht, S. H. and Lawrence, P. S., National survey of congenital malformations resulting from exposure to Roentgen radiation, *Amer. J. Roentgen.*, 73, 442, 1955.

252. Tough, I. M., Buckton, K. E., Kaikie, A. G., and Court-Brown, W. M., X-ray induced chromosome damage in man, *Lancet*, 2, 849, 1960.

253. Buckton, K. E., Jacobs, P. A., Court-Brown, W. M., and Doll, R., A study of the chromosome damage persisting after X-ray therapy for ankylosing spondylitis, *Lancet*, 2, 676, 1962.

254. Goodlin, R. C., Preliminary reports of chromosome studies during radiation therapy, *Amer. J. Roentgen.*, 87, 555, 1962.

255. Millard, R. E., Abnormalities of human chromosomes following therapeutic irradiation, *Cytogenetics*, 4, 277, 1965.

256. Bender, M. A. and Goech, P. C., Persistent chromosome aberrations in irradiated human subjects, *Radiat. Res.*, 16, 44, 1962.

257. Lisco, H. and Conard, R. A., Chromosome studies on Marshall Islanders exposed to fallout radiation, *Science*, 157, 445, 1967.

258. Bloom, A. D., Awa, Akio, Neriishi, Shotaro, and Honda, Take, Chromosome aberrations in leucocytes of older survivors of the atomic bombings of Hiroshima and Nagasaki, *Lancet*, October, 802, 1967.

259. Bloom, A. D., Neriishi, S., and Archer, P. G., Cytogenetics of the *in utero* exposed of Hiroshima and Nagasaki, *Lancet*, 2, 10, 1968.

260. Chu, E. H. Y., Giles, N. H., and Passano, K., Types and frequencies of human chromosome aberrations induced by X-rays, *Proc. Nat. Acad. Sci.*, 47, 830, 1961.

261. Bender, M. A. and Gooch, P. C., Types and rates of X-ray induced chromosome aberrations in human blood irradiated *in vitro*, *Proc. Nat. Acad. Sci.*, 48, 522, 1962.

262. Bell, A. G. and Baker, D. G., Irradiation induced chromosome aberrations in normal human leukocytes in culture, *Canad. J. Genet. Cytol.*, 4, 340, 1962.

263. Bender, M. A., Chromatid-type aberrations induced by X-rays in human leukocyte cultures, *Cytogenetics*, 2, 107, 1963.

264. Dekaban, A. S., Thron, R., and Steusing, J. K., Chromosomal aberrations in irradiated blood and blood cultures of normal subjects and of selected patients with chromosomal abnormality, *Radiat. Res.*, 27, 50, 1966.

265. Schmickel, Roy, Chromosome aberrations in leukocytes exposed *in vitro* to diagnostic levels of X-rays, *Amer. J. Hum. Genet.*, 19, 1, 1967.

266. Sasaki, M., Ottomen, R. E., and Norman A., Radiation induced chromosome aberrations in men, *Radiology*, 81, 652, 1963.

267. Court-Brown, W. M., Buckton, K. E., and McLean, A. S., Quantitative studies of chromosome aberrations in man following acute and chronic exposure to X-rays and gamma rays, *Lancet*, 1, 1239, 1965.

268. Norman, R., Sasaki, M., Ottoman, R. E., and Veemett, R. C., Chromosome aberrations in radiationworkers, *Radiat. Res.*, 23, 282, 1964.

269. Stewart, J. S. S. and Sanderson, A. R., Chromosome aberration after diagnostic X-irradiation, *Lancet*, 1, 978, 1961.

270. Conen, P. E., Chromosome damage in an infant after diagnostic X-irradiation, *Lancet*, 2, 47, 1961.

271. Bloom, A. D. and Tijo, J., *In vivo* effects of diagnostic X-irradiation in peripheral leukocyte chromosomes in man, *Blood*, 22, 822, 1963.

272. Kucerova, Maria, Chromosome analysis of an infant after intrauterine irradiation, *Acta Radiol.*, 6, 410, 1967.

273. Uchida, I. A., Holunga, R., and Lawler, C., Maternal radiation and chromosomal aberrations, *Lancet*, 2, 1045, 1968.

274. Reisman, L. E., Jacobson, A., Davis, L. A., Kasahara, S., and Kelly, S., Effects of diagnostic X-rays on chromosomes in infants. A preliminary report, *Radiology*, 89, 75, 1967.

275. Terry, L. L. and Chadwick, D. R., Current concepts in radiation protection, *J.A.M.A.*, 180, 995, 1962.

276. Pochin, E. E., The development of the quantitative bases for radiation protection, *Brit. J. Radiol.*, 43,155, 1970.

277. Snavely, David R., Kimbier, L. G., Thompson, M. J., and Setter, L. R., Regulations, standards, and guides pertaining to medical and dental radiation protection – An annotated bibliography, U. S. Department of Health, Education and Welfare, Public Health Service Publication #999-RH-37, June 1969.

278. Radiation quantities and units, ICRU Report 10a, National Bureau of Standards, HB 84, 1962.

279. ICRU Report #2, International X-ray unit of intensity, *Brit. J. Radiol.*, 1, 363, 1928.

280. Taylor, Lauriston S., Brief history of the national committee on radiation protection and measurements (NCRP) covering the period 1929-1946, *Health Phys.*, 1, 3, 1958.

281. Boffey, T. M., Radiation standards: Are the right people making decisions?, *Science* 171, 780, 1971.

282. NCRP Report #39,Basic radiation protection criteria, 1971.

283. NCRP Report #4, Radium protection, 1938. *Phys.*, 1, 306, 1958.

284. Sievert, R. M., The tolerance dose and the prevention of injuries caused by ionizing radiation, *Brit. J. Radiol.*, 20, 306, 1947.

285. NCRP Report #17, Permissible dose from external sources of ionizing radiation, National Bureau of Standards, H.B. #59, 1954.

286. Jacobson, L. E., Schwartzman, J. J., and Heiser, S., Monitoring of the diagnostic X-ray department, *Radiology*, 58, 568, 1952.

287. Bushong, S. C., Harle, T. S., and Pogonowska, M. J., Recommendations for personal monitoring in diagnostic radiology, *Phys. Med. Biol.*, 15, 91, 1970.

288. Williams, E. S., Radiation hazard to female radiographers, *Brit. J. Radiol.*, 40, 960, 1967.

289. Langmead, W. A., Radiation hazard to female radiographers, *Brit. J. Radiol.*, 41, 75, 1968.

290. Bushong, S. C., Reduction of occupational exposure in a general purpose radiology room, in preparation.

291. National Academy of Sciences – National Research Council Report, Biological effects of atomic radiation, summary reports, 1956.

292. Taylor, L. S., History of the international commission on radiological protection (ICRP), *Health Phys.* 1, 97, 1958.

293. International Commission on Radiological Protection, Publication #2, Permissible dose for internal radiation, Pergamon Press, 1960.

294. Taylor, L. S., History of the international commission on radiological units and measurements (ICRU), *Health Phys., 1, 306, 1958.*

295. ICRP and ICRU report, Exposure of man to ionizing radiation arising from medical procedures (An inquiry into methods of evaluation), *Phys. Med. Biol.*, 2, 107, 1957.

296. ICRP and ICRU report, Exposure of man to ionizing radiation arising from medical procedures with special reference to radiation induced diseases (An inquiry into methods of evaluation), *Phys. Med. Biol.*, 6, 199, 1961.

297. Taylor, L. S., An early portable radiation survey meter, *Health Phys.*, 13, 1347, 1967.

298. Braestrup, C. B., A stray radiation survey of twenty high voltage Roentgen installations, *Radiology*, 31, 206, 1938.

299. Braestrup, C. B., X-ray protection in diagnostic radiology, *Radiology*, 38, 207, 1942.

300. NCRP Report #33, Medical X-ray and gamma ray protection for energies up to 10 MeV – Equipment design and use, 1968.

301. NCRP Report #26, Medical X-ray protection up to three million volts, 1961.

302. Bell, A. L. L., X-Ray therapy in fluoroscopy, *Radiology*, 40, 139, 1943.

303. Schatzki, R., Medical progress: Diagnostic Roentgenology; Dangers associated with fluoroscopy, *New Eng. J. Med.*, 227, 18, 1942.

304. Morgan, R. H., Protection from Roentgen rays, *Amer. J. Med.*, 226, 578, 1953.

305. Chamberlain, W. E., Fluorscopes and fluoroscopy, *Radiology*, 38, 383, 1942.

306. Sonnenblick, B. P., X-ray exposure in routine diagnostic practice: A survey of 117 fluoroscopes, *Genetics*, 37, 627, 1952.

307. Kirsch, I. E., Radiation dangers in diagnostic radiology, *J.A.M.A.*, 158, 1420, 1955.

308. Gitlin, J. N. and Lawrence, P. S., Population exposure to X-rays, U. S. 1964, Public Health Service Publication #1519, 1966.

309. Ditchek, T., A medical X-ray survey packet – The Surpack, *Health Phys.*, 10, 605, 1964.

310. Frisoli, A., Interpretation of medical X-ray survey packet results, *Health Phys.*, 11, 822, 1965.

311. Ditchek, T., Frisoli, A., and Jones, D., A medical X-ray survey packet, *Health Phys.*, 12, 341, 1966.

312. Chadwick, D. R., Spector, M. I., and Kincaid, C. B., Observations derived from film packs used in a nation-wide X-ray exposure study, *Radiology*, 87, 321, 1966.

313. Fess, L. H. and Seabron, L. C., Preliminary results of 5263 X-ray protection surveys of facilities with medical X-ray equipment (1962-1967), U. S. Public Health Service Publication MORP 68-6, April 1968.

314. Seagle, E. F., X-ray equipment survey in Polk County, Florida, September 1961-August 1963, U. S. Public Health Service Publication #999-RH-8, 1964.

315. Seabron, L. C., Radiation safety surveys of X-ray facilities within the Bureau of Prisons during 1968, U. S. Public Health Service Publication #BRH/DEP, 70-19, 1970.

316. Miller, L. A. and Seabron, L. C., Radiation safety surveys of X-ray facilities within the Federal Health Program Service, U. S. Public Health Service Publication #BRH/DEP 70-17, 1970.

317. ICRP Publication #16, Protection of the Patient in X-ray Diagnosis, 1969.

318. McCullough, E. C. and Cameron, J. R., Exposure rates from diagnostic X-ray units, *Brit. J. Radiol.*, 43, 448, 1970.

319. Jackson, W., Dose assessment in diagnostic radiology, *Brit. J. Radiol.*, 40, 301, 1967.

320. Wilsey, R. B., The use of photographic films for monitoring stray X-rays and gamma rays, *Radiology*, 56, 1951.

321. Gorson, R. O., Suntharalingam, and Thomas, J. W., Results of a film-badge reliability study, *Radiology*, 84, 333, 1965.

322. Barber, donald E., Standards of performance for film badge services, U. S. Department of Health, Education and Welfare, #999-RH-20, 47, 1966.

323. Pocket dosimeters? Film badges?...or Both?, *Nucleonics*, 17, #5, 116, May 1959.

324. Kathren, R. L. and Yoder, R. E., Personnel monitoring with film, glass, and TLD, Western Industrial Health Conference at Los Angeles, 1, October 1966.

325. Cusimano, J. P., Cipperley, F. V., and Culley, J. D., Field experience with termoluminescent dosimeters, Radiations dosimetry based on solid state phenomena school, *NATO*, 1, September 1967.

326. Berstein, Irving A., Bjarngard, Bengt, E., and Jones, Douglas, On the use of phosphor-teflon termoluminescent dosimeters in health physics, *Health Phys.*, 14, 33, 1968.

327. Hall, R. M. and Wright, C. N., A comparison of lithium fluoride and film for personnel dosimetry, *Health Phys.* 14, 37, 1968.

328. Lindblom, A., Measuring of doses received by the personnel in radiological departments and its importance, *Strahlentherapie*, 124, 69, 1961.

329. Macht, S. H. and Katz, E. R., Detection of faulty Roentgenoscopic technique by direct radiation measurements, *Amer. J. Roentgen.*, 68, 809, 1952.

330. Jones, D. E. A., The radiological survey of daignostic X-ray equipment, *Brit. J. Radiol.*, 41, 833, 1968.

331. Code of Practice for the protection of Persons against Ionizing Radiation arising from Medical and Dental Use, HMSO, 1964.

332. U. S. Atomic Energy Commission, News Release, Vol. 1, #22, December 1970.

333. Recommendations to the Board of Directors of the Health Physics Society from the ad hoc committee on radiation exposure records, *Health Phys.*, 14, 157, 1968.

334. Buchan, R. C. T., The limitations of personal radiation monitoring in the radiodiagnostic department, *Brit. J. Radiol.*, 41, 876, 1968.

335. Oliver, R., Medical supervision of radiation workers, *Brit. J. Radiol.*, 43, 497, 1970.

336. Bushong, S. C., Cox, J. L., Collins, V. P., Neibel, J. B., and Murphy, G. B., Eds., Medical radiation information for litigation, U. S. Public Health Service Document, DMRE 69-3, July 1969.

337. Ellis, R., Medical supervision of radiation workers, *Brit. J. Radiol.*, 43, 498, 1970.

338. Hill, S. M. B., Medical supervision of radiological workers, *Brit. J. Radiol.*, 43, 751, 1970.

339. ICRP Publication #9, Radiation protection, 1965.

340. Oliver, R., Are dose records and medical supervision necessary for low level radiation exposures?, *Brit. J. Radiol.*, 42, 397, 1969.

341. Henshaw, E. T., Cumulative radiation dose records, *Brit. J. Radiol.*, 42, 716, 1969.

342. Doust, C. E. and STern, B. E., Radiation dose records, *Brit. J. Radiol.*, 41, 240, 1968.

343. Poppel, M. H., Sorrentino, J., and Jacobson, H. G., Personal diary of radiation dosage: Plea for standardized system, *J.A.M.A.*, 147, 630, 1951.

344. Hodges, P. C., Health Hazards in the diagnostic use of X-ray,*J.A.M.A.*, 166, 577, 1958.

345. Recommended safe practice in the use of diagnostic X-ray equipment, Industrial Medical Association, Chicago, 1963.

346. Buchan, R. C. T., Protective aprons and film badges, *Brit. J. Radiol.*, 40, 73, 1967.

347. Buchan, R. C. T., Film badges and protective aprons, *Brit. J. Radiol.*, 40, 238, 1967.

348. Jones, D. E. A., Film-badge monitoring in X-ray departments, *Brit. J. Radiol.*, 40, 73, 1967.

349. Bushong, S. C., Radiation control in diagnostic Roentgenology, *Health Phys.*, 19, 557, 1970.

350. Leddy, Eugene T., The causes of Roentgen ray dermatitis among physicians, *Amer. J. Roentgen.*, 36, 510, 1936.

351. Stevenson, Clyde A., and Leddy, Eugene T., The dangers of reducing fractures under the Roentgenoscope and methods of protection against them, *Amer. J. Roentgen.*, 37, 70, 1937.

352. Leddy, E. T., Roentgenologic risks sustained by physician not trained in Roentgenology, *Med. Clin. N. Amer.*, July 10-11, 1941.

353. Harding, D. B., Radiation hazards in medical practice, *Kentucy Med. J.*, 43, 228, 1945.

354. Ritvo, M., D'Angio, G. J., and Rhodes, I. E., Radiation hazards to nonradiologists participating in X-ray examinations, *J.A.M.A.*, 160, 4, 1956.

355. Nagle, Robert B. and Peirson, Edward L., A study of the radiation hazard in urology, *J. Urol.*, 70, 338, 1953.

356. Rytila, A. and Perttala, Y., Average doses to the radiologist in contrast-media studies of the gastrointestinal tract, lower extremity venography and urethrocystography, *Health Phys.*, 18, 123, 1970.

357. Dubilier, William, Jr., Burnett, Harry W., Dotter, Charles T., and Steinberg, Israel, Radiation hazard during angiocardiography, *Amer. J. Roentgen.*, 70, 441, 1953.

358. Hills, T. H. and Stanford, R. W., The problem of excessive radiation during routine investigations of the heart, *Brit. Heart J.*, 12, 45, 1950.

359. Liden, Kurt and Lindgren, Martin, Radiation hazards during cholangiographic examinations, *Acta Radiol.*, 38, 1, 1952.

360. Cowing, R. F. and Spalding, C. K., Survey of scattered radiation from fluoroscopic units in fifteen institutions, 53, 569, 1949.

361. Soiland, A., Protection to the operator from unnecessary radium radiation, *Amer. J. Roentgen.*, IX, 683, 1922.

362. Nuttall, J. R., A scheme of protection which has proved satisfactory in a large radium therapy center, *Amer. J. Roentgen.*, 41, 98, 1939.

363. Wilson, C. W., Protection in radium therapy at Westminster Hospital: A summary of measurements of gamma-ray doses made over a number of years with condenser ionization chambers carried by the staff, *Brit. J. Radiol.*, 13, 105, 1940.

364. Wilson, C. W. and Greening, J. R., Gamma ray protection in radium therapy, *Brit. J. Radiol.*, 21, 211, 1948.

365. Taft, Robert B., Stray radiation under actual conditions, *Amer. J. Roentgen.*, 46, 373, 1941.

366. Fleming, J. A. C., Investigations into the degree of scattered radiation received by X-ray workers during routine diagnostic examinations in a military hospital department, *Brit. J. Radiol.*, 16, 367, 1943.

367. Binks, W., Radiation exposure of staff in diagnostic procedures, Part III, Some aspects of radiation hygiene, *Brit. J. Radiol.*, 28, 654, 1955.

368. Braestrup, C. B., Past and present radiation exposure to radiologists from the point of view of life expectancy, *Amer. J. Roentgen.*, 78, 988, 1957.

369. Clark, L. H. and Jones, D. E. A., Some results of the photographic estimation of stray X-radiation received by hospital X-ray personnel, *Brit. J. Radiol.*, 16, 166, 1943.

370. Cowie, Dean B. and Scheele, Leonard A., A survey of radiation protection in hospitals, *J. Nat. Cancer Inst.*, I, 767, 1940-1941.

371. DeAmicis, Egilda, Spalding, Charles K., and Cowing, Russel F., Survey of X-ray exposures in hospital personnel, *J.A.M.A.*, 149, 924, 1952.

372. Geist, Robert M., Jr., Glasser, Otto, Hughes, C., Robert, Radiation exposure survey of personnel at the Cleveland Clinic Foundation, *Radiology*, 60, 186, 1953.

373. Heustis, Albert E. and Van Farowe, Donald, The hazard of radiation, *Radiation*, 57, 832, 1951.

374. Osborn, S. B., Radiation doses received by diagnostic X-ray workers, Part II, *Brit. J. Radiol.*, 28, 650, 1955.

375. Jamieson, H. D., X-ray dosage to patients and staff in diagnostic radiology (An investigation at Dunedin Hospitals), *New Zeal. Med. J.*, 51, 159, 1952.

376. Cowing, R. F., Radiation dosage to medical personnel, *Amer. Industr. Hyg. J.*, 21, 169, 1960.

377. Spalding, C. K. and Cowing, R. F., A summary of radiation exposures received by workers in medical X-ray departments from 1950-1960, *Health Phys.*, 8, 499, 1962.

378. Langmead, Walter A. and Steadman, E. J., Radiation doses received by staff in British hospitals and their relation to maximum permissible dose, *Brit. J. Radiol.*, 43, 279, 1970.

379. Duggan, W. J., Greenslade, R., and Jones, D. E., External radiation doses from occupational exposure, *Nature*, 221, 831, 1969.

380 Malsky, S. J., The use of radio TLD in the field of diagnostic radiology, *Radiology*, 90, 518, 1968.

381. Attix, F., Glass and thermoluminescent dosimetry in *Technological Needs for Reduction of Patient Dosage from Diagnostic Radiology*, Janower, M. L., Ed., Charles C Thomas, Springfield, Ill., 1963, 61.

382. Wykoff, H., Quantities and units for dosimetry in diagnositc radiology, in *Technological Needs for Reduction of Patient Dosage from Diagnostic Radiology*, Janower, M. L., Ed., Charles C Thomas, Springfield, Ill., 1963, 5.

383. Morgan, Russell H. and Gehret, Judith C., The radiant energy received by patients in diagnostic X-ray practice, *Amer. J. Roentgen*, 97, 793, 1966.

384. Miller, E. R., Kufafian, J., and Maddison, F. E., Patient exposure during fluoroscopy, *Radiology*, 80, 477, 1963.

385. Ardran, G. M. and Crooks, H. E., The measurement of patient dose, *Brit. J. Radiol.*, 38, 766, 1965.

386. Pychlau, p. and Bunde, E., The absorption of X-rays in a body equivalent phantom, *Brit. J. Radio.*, 38, 875, 1965.

387. Pychlau, P., Dose assessment in diagnostic radiology, *Brit. J. Radiol.*, 40, 559, 1967.

388. Sprawls, P., Miller, W. B., Weens, H. S., and Casper, J. H., Patient X-ray exposure during fluoroscopic examinations with an analysis of contributing parameters, *Radiology*, 87, 99, 1966.

389. Carlson, Carl, Integral absorbed doses in Roentgen diagnostic procedures, *Acta Radiol.*, 3, 384, 1965.

390. Ardran, G. M., Mamill, J., Emrys Robert E., and Olifer, R., Radiation dose to the patient in cardiac radiology, *Brit. J. Radiol.*, 43, 391, 1970.

391. Cameron, J. R., A proposed unit for patient radiation exposure from diagnostic X-rays, *Health Phys.*, in press.

392. Bojtor, Ivan, Bogdany, Barna, and Koczkas, Gyula, Diagnostic film dosimeter in the study of radiation exposure to the population resulting from X-ray fluoroscopy, *Brit. J. Radiol.*, 41, 147, 1968.

393. Koczkas, G. and Bojtor, I., Film dosimetry in radiodiagnosis to assess radiation burden to the population, *Brit. J. Radiol.*, 38, 624, 1965.

394. Rogers, R. T., Radiation dose to the skin in diagnostic radiography, *Brit. J. Radiol.*, 42, 511, 1969.

395. Gough, J. H., Davis, R., and Stacey, A. J., Radiation doses delivered to the skin, bone marrow and gonads of patients during cardiac catheterization and angiocardiography, *Brit. J. Radiol.*, 41, 508, 1968.

396. Radiological Hazards to Patients: Final Report, Ministry of Health, HMSO, 1966.

397. Liuzzi, A., Blatz, H., and Eisenbud, M., A method for estimating the average bone marrow dose from some fluorscopic examinations, *Radiology*, 82, 99, 1964.

398. Yoshinaga, H., Takeshita, K., Sawada, S., Russell, W. J., and Shigetoshi, A., et al., Estimation of exposure pattern and bone marrow, *Brit. J. Radiol.*, 40, 344, 1967.

399. Egan, R. L., Experience with mammography in a tumor institution: Evaluation of 1,000 studies, *Radiology*, 75, 894, 1960.

400. Stanton, L., Lightfoot, D. A., Boyle, J. J., and Cullinan, J. E., Physical aspects of breast radiography, *Radiology*, 81, 1, 1963.

401. Ewtor, J. R., Shalek, R. J., and Egan, R. L., Estimated radiation dose during mammography, *Cancer Bull.*, 14, 116, 1962.

402. Stanton, L. and Lightfoot, D. A., The selection of optimum mammography technic, *Radiology*, 83, 442, 1964. 403.

M. G., and Fingerhut, Aaron G., Evaluation of Roentgen exposure in mammography, Radiology, 95, 395, 1970.

404. Gilbertson, James D., Randall, M. G., and Fingerhut, A. G., Evaluation of Roentgen exposure in mammography, *Radiology*, 97, 641, 1970.

405. Palmer, R. C., Egan, R. L., and Barrett, B. J., Preliminary evaluation of absorbed dose in mammography, *Radiology*, 95, 395, 1970.

406. Heller, M. B., Terdiman, J. F., and Pasternack, B. S., A procedure of calculation of gonadal X-ray dose in diagnostic radiography, *Brit. J. Radiol.*, 39, 686, 1966.

407. Matthews, Jennifer C. and Miller, H., Radiation hazards from diagnostic radiology. A repeat survey over a small area, *Brit. J. Radiol.*, 42, 814, 1969.

408. Stanford, R. W. and Vance, J., The quantity of radiation received by the reproductive organs of patients during routine diagnostic X-ray examinations, *Brit. J. Radiol.*, 28, 266, 1955.

409. Hammer-Jacobson, E., Genetically significant radiation doses in diagnostic radiology, *Acta Radiol.*, Suppl. 222, 1, 1963.

410. Laughlin, J. S. and Pullman, I., *Nat. Acad. Sci.*, Washington, D. C., 1957.

411. Larsson, L. E., Radiation doses to the gonads of patients in Swedish Roentgen diagnostics, *Acta Radiol.*, Suppl. 157, 1, 1958.

412. Radiological Hazards to Patients: Second Report, Ministry of Health, HMSO, London, 1960.

413. McEwan, A. C., Integral and gonad radiation doses in radioisotope renogram and radiographic intravenous pyelogram examinations, *Australas. Radiol.*, 10, 55, 1966.

414. Niki, I., The effect of X-ray quality and ovarian location on genetic exposure in gynecological examinations, *Brit. J. Radiol.*, 39, 607, 1966.

415. Niki, I., A demountable phantom of variable geometry for intra-pelvic dosimetry in pregnancy – preliminary report, *Brit. J. Radiol.*, 38, 472, 1965.

416. Aspin, N., The gonadal x-ray dose to children from diagnostic radiographic technics, *Radiology*, 85, 944, 1965.

417. Kaude, J. V., Lorenz, E., and Reed, J. M., Gonad dose to children in voiding urethrocystography performed with 70 mm image-intensifier fluorography, *Radiology*, 92, 771, 1969.

418. Jackson, Herbert L., Hass, A. Curtis, Sooby, Donna, and Marschke, Charles H., The gonadal exposure of boys and young men treated with inverted "Y" fields: Its reduction and genetic significance, *Radiology*, 96, 181, 1970.

419. Green, James D. and Bushong, Stewart C., Gonadal dose in male radiotherapy patients, *Radiology*, 98, #3, 661, 1971.

420. Hemphill, F. M., Locke, F. B., and Hesselgren, R. D., Diagnostic radiation utilization in selected short-term general hospitals, U. S. Department of Health, Education & Welfare, Public Health Service, *Bureau of Radiological Health Publication* #BRH/DBE 70-8, December 1970, 52.

421. Summary of population exposure to X-rays in the United States, *Radiol. Health Data Rep.*, 8, 367, July 1967.

422. Brown, M. L., Population dose from X-rays, U. S. 1964, *U. S. Public Health Service*, #2001, 82, 1969.

423. Mahmoud, K. A., Mahfouz, M. M., Ativah, I. R., El-Naggar, A. M., and Molokhia, M. M., Genetically significant dose from diagnostic radiology in Cairo and Alexandria, *Health Phys.*, 13, 163, 1967.

424. Izenstark, Joseph L. and Lafferty, W., Medical radiological practice in New Orleans, Estimates and characteristics of visits, examinations and genetically S. D., *Radiology*, 90, 229, 1968.

425. Pasternack, Bernard S. and Heller, M. B., GSD to the population of New York City from diagnostic medical radiology. A dosimetric and statistical study, *Radiology*, 90, 217, 1968.

426. Radiological Hazards to Patients, Interim Report, HMSO, 1959.

427. Penfil, R. L. and Brown, M. L., Genetically significant dose to the United States population from diagnostic medical Roentgenology, 1964, *Radiology*, 90, 209, 1968.

428. Williamson, B. D. P. and McEwan, A. C., The genetically significant dose to the population of New Zealand from diagnostic radiology, Publication NRL/PDS/1965, *Dept. of Health*, Christchurch, New Zealand.

429. Cooley, R. N. and Beentjes, L. B., Weighted gonadal Roentgen ray doses in a teaching hospital with comments in X-ray dosages to the general population of the United States, *Amer. J. Roentgen.*, 92, 404, 1964.

430. Committee Report, The genetically significant dose by the X-ray diagnostic examinations in Japan, *Nippon Acta Radiol.*, 21, 565, 1961.

431. Norwood, W. D., Healy, J. W., Donaldson, E. E., Roesch, W. C., and Kirklin, C. W., The gonadal radiation dose received by the people of a small American city due to diagnostic use of Roentgen rays, *Amer. J. Roentgen.*, 82, 1081, 1959.

432. Biagini, C., Borilla, M., and Montonina, A, An inquiry into exposure of the population of Rome to X-radiation from medical procedures, *ICRP/ICRU*, 1960.

433. Brown, R. F., Heslep, J., and Eads, W., Number and distribution of Roentgenologic examination for 100,000 people, *Radiology*, 74, 353, 1960.

434. Martin, J. H., The contribution to the gene material of the population from the medical use of ionizing radiations, *Med. J. Aust.*, 2, 79, 1958.

435. Westing, S. W., Importance of a new X-ray effect for our daily diagnostic and therapeutic X-ray work, *N. Y. State J. Med.*, 40, 1139, 1940.

436. Report of the Medical X-Ray Advisory Committee on Public Health Considerations in Medical Diagnostic Radiology (X-RAYS), U. S. Department of Health, Education & Welfare, Public Health Service, October 1967.

437. DeLorimier, A. A., Cowie, D. B., and White, T. N., Protective features provided with the United States Army field Roentgenoscopic equipment, *Amer. J. Roentgen.*, 49, 653, 1943.

438. Trout, E. E., Kelley, J. P., and Cathy, G. A., Use of filters to control radiation exposure to patient in diagnostic Roentgenology, *Amer. J. Roentgen.* 67, 946, 1952.

439. Trout, E. D., Kelley, J. P., and Lucas, A. C., The effect of kilovoltage and filtration on depth dose, in *Technological Needs for Reduction of Patient Dosage from Diagnostic Radiology,* Janower, M. L., Ed., Charles C Thomas, Springield, Ill., 1963, 143.

440. Rollins, W., X-light kills, *Boston Med. Surg. J.*, 144, 173, 1901.

441. Hale, J., The effect of collimation: Field size, cone leaks and scatter dose outside the useful beam, in *Technological Needs for Reduction of Patient Dosage from Diagnostic Radiology*, Janower, M. L., Ed., Charles C Thomas, Springfield, Ill., 1963, 159.

442. Goldman, H. L. and Davies, S., Radiation glo-bar: A device for the delineation of a primary X-ray beam, *Health Phys.*, 4, 178, 1960.

443. Blatz, Hanson and Eure, J. A., New York City radiographic cone labels, *Health Phys.*, 12, 1630, 1966.

444. Feldman, A., Babcock, G. C., Lanier, R. R., and Morkovin, D., Gonadal exposure dose from diagnostic X-ray procedures, *Radiology*, 71,197, 1958.

445. Goldman, H. L. and Shultz, H. H., A simple method for reducing the radiation hazard associated with photofluorography, *Amer. Rev. Tuber.*, 77, 923, 1958.

446. Warrick, C. K. and Forster, E., A protection shield for use in chest radiography of children, *Brit. J. Radiol.*, 32, 66, 1959.

447. Ardran, G. M. and Crooks, H. E., Limitation of the primary beam in the chest radiography, *Radiography*, 23, 235, 1956.

448. Gasque, M. R., Shielding device to protect gonads during routine chest Roentgenography, *Industr. Med. Surg.* 27, 79, 1958.

449. Nial, J. W., An abdomen shield, *Radiography*, 24, 17, 1958.

450. Dunning, J. M., A shield for patient protection in dental X-ray work, *Industr. Med. Surg.*, 29, 112, 1960.

451. Steggles, R. W., Fluoroscope apron in gonad protection, *Radiography*, 24, 261, 1958.

452. Barr, M. D., Protection of the abdomen during chest radiography, *Radiography*, 24, 155, 1958.

453. Schwarz, E., Pretto, J. I., and Martin, S., A universal gonadal shield, *Illinois Med. J.*, 118, 24, 1960.

454. Hodges, P. C., Stranjord, N. M., and McCrea, A., A testicular shield, *Illinois Med. J.*, 118, 24, 1960.

455. Magnusson, W., A device for the protection of the testicle in Roentgen examinations of adjacent organs and bones, *Acta Radiol.*, 37, 288, 1952.

456. Ardran, G. M. and Kemp, F. H., Protection of the male gonads in diagnostic procedures, *Brit. J. Radiol.*, 30, 280, 1957.

457. Steadman, E. J., Radiation hazards from diagnostic radiology, *Brit. J. Radiol.*, 42, 946, 1969.

458. Smith, Roy H. and Willhoit, D. G., Dose reduction factors of a leadlined girdle, *Health Phys.*, 19, 380, 1970.

459. Cassady, J. R. and Pierce, R. N., A study of head and neck exposure from polytomography, *U. S. Public Health Service Publication BRH/DEP 70-27*, November 1970.

460. Chin, F. K., Anderson, W. B., and Gilbertson, J. D., Radiation dose to critical organs during petrous tomography, *Radiology*, 94, 623, 1970.

461. Janower, M. L., Ed., *Technological Needs for Reduction of Patient Dosage from Diagnostic Radiology*, Charles C Thomas, Springfield, Ill., 1963.

462. Proceedings of the Symposium on the Reduction of Radiation Dose in Diagnostic X-Ray Procedures, July 8, 1971, in press.

463. Bates, Lloyd, Some physical factors affecting radiographic image quality, their theoretical basis and measurement, U. S. Department of Health, Education and Welfare, *Public Health Service Publication #999-RH-38*, August 1969.

464. Goldberg, R. M., A simple means for reducing unnecessary patient dose in radiology departments, *Health Phys.*, 19, 311, 1970.

465. Properzio, W. S. and Trout, E. D., The deterioration of X-ray fluoroscopic screens, *Radiology*, 91, 439, 1968.

466. Docker, M. and Astley, R., A video tape recorder used to reduce patient dose during X-ray screening, *Brit. J. Radiol.*, 42, 358, 1969.

467. Ardran, G. M. and Crooks, H. E., Dose reduction in fluoroscopy, *Brit. J. Radiol.*, 42, 554, 1969.

468. Astley, R. and Docker, M. F., Dose reduction in fluoroscopy, *Brit. J. Radiol.*, 42, 635, 1969.

469. White, T. N., Cowie, D. B., and Delosmer, A. A., Radiation hazards during Roentgenoscopy, *Amer. J. Roentgen.*, 19, 639, 1943.

470. Engeset, Arne, A new protector in cerebral angiography, *Acta Radiol.*, 29, 503, 1948.

471. Ross, J. A., Detachable protection apron, *Brit. J. Radiol.*, 40, 74, 1967.

472. Hranitsky, E. B. and Shalek, R. J., Skin doses from glass and beryllium window X-ray tubes in mammography, *Radiology*, 88, 668, 1967.

473. Maudal, Sem, Iron filters as a means of reducing the dose in Roentgen examination of the female breast, *Acta Radiol.*, 7, 239, 1967.

474. Price, J. L. and Butler, P. D., the reduction of radiation and exposure time im mammography, *Brit. J. Radiol.*, 43, 251, 1970.

475. Berridge, F. R., The requirements in design of an X-ray table for fluoroscopic examinations, *Brit. J. Radiol.*, 44, 1950, 1971.

476. Oliver, R., Radiation exposure in diagnostic X-ray departments, *Brit. J. Radiol.*, 42, 633, 1969.

477. Bohrer, S. P., Radiation exposure in diagnostic X-ray departments, *Brit. J. Radiol.*, 43, 429, 1970.

478. Blatz, Hanson, Laboratory support for an X-ray inspection program, *Health Phys.*, 10, 643, 1964.

479. Radiation Protection Survey Report Manual, National Center for Radiological Health, *U. S. Public Health Service*, January 1967.

480. Chamberlain, R. H., Medical radiation: Perspectives in risks and rewards, *Amer. J. Public Health*, 55, 710, 1965.

481. Niki, I., Exposure to the maternal ovaries resulting from pregnancy radiographs as a function of ovarian location, *Brit. J. Radiol.*, 40, 453, 1967.

482. Brown, M. L., Roney, P. L., Gitlin, J. N., and Moore, R. T., X-Ray experience during pregnancy, *J.A.M.A.*, 199, 87, 1967.

483. Jacobsen, L., Experimental and practical efforts to minimize the radiation induced hazards in diagnostic departments, *Brit. J. Radiol.*, 42, 154, 1969.

484. Reekie, D., Davison, M., and Davidson, J. L., The radiation hazard in radiography of the female abdomen and pelvis, *Brit. J. Radiol.*, 40, 849, 1967.

485. Langmead, W. A., Diagnostic radiology in relation to the menstrual cycle, *Brit. J. Radiol.*, 41, 952, 1968.

486. Ellis, Frank, Diagnostic radiology in relation to the menstrual cycle, *Brit. J. Radiol.*, 41, 877, 1968.

487. Bennett, Robert, Diagnostic radiology in relation to the menstrual cycle, *Brit. J. Radiol.*, 41, 952, 1968.

488. Bennett, R., Some aspects of radiation protection in a diagnostic department. Emphasizing elective examination of females of reproductive age, *Australas. Radiol.*, 12, 224, 1969.

489. Stieve, F. E., Estimation of the radiation dose in diagnostic radiology, *Radiography*, 31, 173, 1965.

490. Kinlen, L. J. and Acheson, Ed, Diagnostic irradiation, congenital malformations and spontaneous abortion, *Brit. J. Radiol.*, 41, 648, 1968.

491. Bowerman, J. W., El-Mahdi, A. M., Lott, J. S., and Julian, C. G., Radiation abortion in radiotherapy, *Radiology*, 91, 1013, 1968.

492. Hammer-Jacobsen, E., Therapeutic abortion on account of X-ray examination during pregnancy, *Danish Med. Bull.*, 6, 133, 1959.

493. Wolf, B. S. and Greenberg, E. I., Radiation, pregnancy, and progency, in *Medical Surgical, and Gynecologic Complications of Pregnancy*, Ravinsky, J. J., Ed., Williams & Co., Baltimore, 1960.

494. Field, C. E., Carcinoma of the uterus with pregnancy intervening treated successfully by radium followed by delivery of a normal child, *Amer. J. Roentgen*, 657.

495. Adams, F. H. and Rigler, L. G., Reduction of radiation to children, *Circulation*, 30, 161, 1964.

496. Dawson, J., Phillips, G. M., and Willcox, B. A. M., Computation of X-ray diagnostic exposure tables, *Brit. J. Radiol.*, 39, 117, 1966.

497. Lodwick, G. S., Recommendations for obtaining the maximum benefit of radiation exposure in diagnostic radiology through improved production and utilization of image information, Report to the Natl. Center for Radiol. Health, U.S. Public Health Service Contract #PH86-67-198, 1, July, 1968.

498. Speigler, G. and Keane, B. E., Image contrast and radiation protection: A figure of merit, Part I-Narrow beams, *Brit. J. Radiol.*, 38, 771. 1965.

499. Speigler, G. and Keane, B. E., Image contrast and radiation protection: A figure of merit, Part II-Wide beams, *Brit. J. Radiol.*, 38, 871, 1965.

500. Morgan, R. H. and Chaney, H. E., Darkroom practice and unnecessary patient exposure, *Amer. J. Roentgen*, 94, 236, 1965.

501. Dockray, K. T., Timed work steps during diagnostic radiography, *Radiology*, 91, 497, 1968.

502. Birnkrant, M. I. and Henshaw, P. S., Further problems in X-ray protection. I. Radiation hazards in photofluorography, *Radiology*, 44, 565, 1945.

503. Morgan, R. H. and Lewis, I., The protection of photofluorographic personnel, *Amer. J. Roentgen.*, 55, 198, 1946.

504. Tizes, R. and Tizes, C. W., Decline in statewide mobile X-ray programs to detect tuberculosis, *Public Health Rep.*, 85, 901, 1970.

505. Proceedings of the National Conference on X-Ray Technician Training, Division of Radiological Health, U. S. Public Health Services, September 7-9, 1966.

506. S. 426, Radiation Health and Safety Act of 1971.

507. The Control of Radiation Hazards in the United States, Report of the National Advisory Committee on Radiation, 1959.

508. Protecting and Improving Health through the Radiological Sciences, Report of the National Advisory Committee on Radiation, 1966.

509. Report on a Study of Academic Radiology, National Academy of Sciences, National Research Council, 1969.

510. Brown, R. F., reported in *U. S. Public Health Service BRH Bulletin 5:2*, May 10, 1971.

INDEX

I

L

M

N

O

P

R

T

U